SI units

Physical quantity	*Old unit*	*Value in SI units*
energy	calorie (thermochemical)	4·184 J (joule)
	*electronvolt—eV	$1{\cdot}602 \times 10^{-19}$ J
	*electronvolt per molecule	96·48 kJ mol^{-1}
	erg	10^{-7} J
	*wave number—cm^{-1}	$1{\cdot}986 \times 10^{-23}$ J
entropy (S)	eu = cal g^{-1} $°C^{-1}$	4184 J kg^{-1} K^{-1}
force	dyne	10^{-5} N (newton)
pressure (P)	atmosphere	$1{\cdot}013 \times 10^{5}$ Pa (pascal), or N m^{-2}
	torr = mmHg	133·3 Pa
dipole moment (μ)	debye—D	$3{\cdot}334 \times 10^{-30}$ C m
magnetic flux density (H)	*gauss—G	10^{-4} T (tesla)
frequency (ν)	cycle per second	1 Hz (hertz)
relative permittivity (ε)	dielectric constant	1
temperature (T)	*°C and °K	1 K (kelvin); 0 °C = 273·2 K

(* indicates permitted non-SI unit)

Multiples of the base units are illustrated by length

fraction	10^{9}	10^{6}	10^{3}	1	(10^{-2})	10^{-3}	10^{-6}	10^{-9}	(10^{-10})	10^{-12}
prefix	giga-	mega-	kilo-	metre	(centi-)	milli-	micro-	nano-	(*ångstrom)	pico-
unit	Gm	Mm	km	m	(cm)	mm	μm	nm	(*Å)	pm

The fundamental constants

Avogadro constant	L or N_A	$6{\cdot}022 \times 10^{23}$ mol^{-1}
Bohr magneton	μ_B	$9{\cdot}274 \times 10^{-24}$ J T^{-1}
Bohr radius	a_0	$5{\cdot}292 \times 10^{-11}$ m
Boltzmann constant	k	$1{\cdot}381 \times 10^{-23}$ J K^{-1}
charge of a proton (charge of an electron = $-e$)	e	$1{\cdot}602 \times 10^{-19}$ C
Faraday constant	F	$9{\cdot}649 \times 10^{4}$ C mol^{-1}
gas constant	R	8·314 J K^{-1} mol^{-1}
nuclear magneton	μ_N	$5{\cdot}051 \times 10^{-27}$ J T^{-1}
permeability of a vacuum	μ_0	$4\pi \times 10^{-7}$ H m^{-1} or N A^{-2}
permittivity of a vacuum	ε_0	$8{\cdot}854 \times 10^{-12}$ F m^{-1}
Planck constant	h	$6{\cdot}626 \times 10^{-34}$ J s
(Planck constant)/2π	$\hbar$	$1{\cdot}055 \times 10^{-34}$ J s
rest mass of electron	m_e	$9{\cdot}110 \times 10^{-31}$ kg
rest mass of proton	m_p	$1{\cdot}673 \times 10^{-27}$ kg
speed of light in a vacuum	c	$2{\cdot}998 \times 10^{8}$ m s^{-1}

$\ln 10 = 2{\cdot}303$ $\ln x = 2{\cdot}303 \lg x$ $\lg e = 0{\cdot}4343$ $\pi = 3{\cdot}142$

$R \ln 10 = 19{\cdot}14$ J K^{-1} mol^{-1} $RTF^{-1} \ln 10 = 59{\cdot}16$ mV at 298·2 K

Oxford Chemistry Series

Oxford Chemistry Series

1972

1. K. A. McLauchlan: *Magnetic resonance*
2. J. Robbins: *Ions in solution* (2)*: an introduction to electrochemistry*
3. R. J. Puddephatt: *The periodic table of the elements*
4. R. A. Jackson: *Mechanism: an introduction to the study of organic reactions*

1973

5. D. Whittaker: *Stereochemistry and mechanism*
6. G. Hughes: *Radiation chemistry*
7. G. Pass: *Ions in solution* (3)*: inorganic properties*
8. E. B. Smith: *Basic chemical thermodynamics*
9. C. A. Coulson: *The shape and structure of molecules*
10. J. Wormald: *Diffraction methods*
11. J. Shorter: *Correlation analysis in organic chemistry: an introduction to linear free-energy relationships*
12. E. S. Stern (ed.): *The chemist in industry* (1)*: fine chemicals for polymers*
13. A. Earnshaw and T. J. Harrington: *The chemistry of the transition elements*

JOHN SHORTER
READER IN PHYSICAL ORGANIC CHEMISTRY,
UNIVERSITY OF HULL

Correlation analysis in organic chemistry:

an introduction to linear free-energy relationships

Clarendon Press · Oxford · 1973

Oxford University Press, Ely House, London W.1
GLASGOW NEW YORK TORONTO MELBOURNE WELLINGTON
CAPE TOWN IBADAN NAIROBI DAR ES SALAAM LUSAKA ADDIS ABABA
DELHI BOMBAY CALCUTTA MADRAS KARACHI LAHORE DACCA
KUALA LUMPUR SINGAPORE HONG KONG TOKYO

PHOTOTYPESET BY
OLIVER BURRIDGE FILMSETTING LTD, CRAWLEY, SUSSEX
PRINTED IN GREAT BRITAIN BY
J. W. ARROWSMITH LTD, BRISTOL, ENGLAND

Editor's foreword

IN the evolution of any scientific discipline the correlation of apparently isolated items of information is a necessary prerequisite to the formulation of an underlying theory. In 1937 Hammett evolved a semi-empirical correlation of reaction rates for side-chain reactions in *meta*- and *para*-substituted aromatic compounds. This approach to linear free-energy relationships has formed the basis of many subsequent developments. Today, this type of treatment is used in such widely differing areas as correlations of n.m.r. chemical-shift values on the one hand to structure–activity relationships of drugs on the other. Clearly, the field is of interest to a wide range of chemists.

The subject of linear free-energy relationships has perhaps not occupied as prominent a place in undergraduate courses as its importance would merit. One of the reasons for this may well have been the absence of a suitable undergraduate text. This book fills the gap and will undoubtedly form the basis for undergraduate courses in this area in the future.

The author, John Shorter, is not only a prominent researcher in the field but has also been a co-editor of the standard research monograph, *Advances in linear free-energy relationships*, Plenum Press, 1972. Since he is an experienced university chemistry teacher it is clear that nobody is better equipped to produce an undergraduate text in this area. The book is suitable for the final year of an Honours course in chemistry and will be of interest to many research workers in widely differing fields of chemical endeavour.

J.S.E.H.

Preface

AT the time when I was invited to write this book, Professor N. B. Chapman and I were editing *Advances in linear free-energy relationships*, Plenum Press, 1972. The latter is a multi-author international research monograph. This editorial work provided the immediate background to my writing the present book in which I have tried to deal with the subject at a level suitable for a final-year undergraduate or first year post-graduate student. I have set out to show the variety of applications of correlation analysis to a wide range of chemical and related sciences.

I assume that readers have a basic knowledge of organic chemistry: nomenclature, characteristic reactions of the common functional groups, reaction mechanisms, and electronic theory. A good grounding in physical chemistry is also assumed—valency theory, kinetics, equilibria, and spectroscopy—but brief introductory accounts of some topics are given where these may be helpful. No previous knowledge of enzymology or pharmacology is assumed for Chapter 7, and the relevant topics in statistics are outlined in an Appendix.

Various friends have assisted me. The contributors to *Advances in linear free-energy relationships* provided me with a stimulating background; individual acknowledgements appear in the Bibliography. Professor Chapman and I have collaborated in research and writing for sixteen years and I am glad to acknowledge his constant encouragement, helpful advice, and sound criticism; he has read the whole manuscript. Several colleagues in Hull have given expert comment on certain parts of the book: Dr. G. W. Crosbie, Dr. D. F. Ewing, Dr. R. B. Moyes, and Mr. G. Collier. The Appendix has been criticized by Professor R. J. Nicholson of the Division of Economic Studies at Sheffield, and formerly of Hull. For many years he has tried to keep me on the right lines in statistics and I am greatly indebted to him. Mr. R. Wilkinson, my research technician, has done computational work for some of the problems.

I am grateful to Mrs. J. Naylor (Supervisor of the University typing pool), and to Miss Lynne Taylor and Miss Jennifer Trowman of the Chemistry Department for their careful production of the typescript, to the Series Editors, and to the staff of the Clarendon Press. Finally, thanks are due to my wife and family for enduring yet another period in which I have somewhat neglected domestic concerns.

The University, Hull
September 1972

JOHN SHORTER

Contents

1. Introduction and general survey

What is 'Correlation Analysis'?

DURING the last half century many rate constants and equilibrium constants for organic reactions in solution have been measured. The total volume of results available is so large that it is vital to have suitable procedures for summarizing and analysing them. The summarizing involves the development and use of *empirical correlations*, whereby one body of results can be related to another. The data may thus also be analysed to reveal the fundamental factors underlying organic reactivity. This general approach is called 'Correlation Analysis'. It is important not only for the qualitative explanation of organic reactivity, which has long been practised, but also for the more quantitative aims of theoretical chemistry.

Empirical correlations of the reactivities of organic compounds are usually linear relationships involving logarithms of rate (k) or equilibrium (K) constants. An elementary example of such a relationship is shown in Fig. 1,

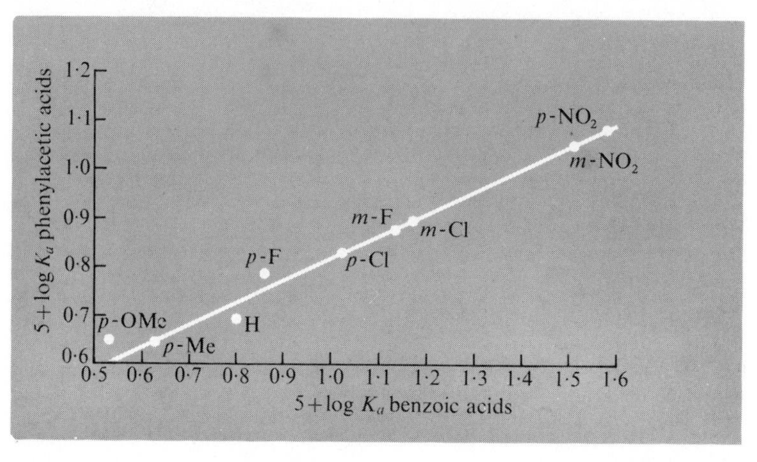

FIG. 1. Linear free-energy relationship for ionization of substituted phenylacetic and benzoic acids. K_a values (water, 25°C): phenylacetic acids, Fischer, Mann, and Vaughan (1961); benzoic acids, Dippy (1939).

in which log K_a values for the ionization of ring-substituted phenylacetic acids **1** are plotted against log K_a values for the ionization of the corresponding substituted benzoic acids **2**. Such a correlation may be described as an

CH_2CO_2H X **1** CO_2H X **2**

equilibrium–equilibrium relationship; we may also have rate–rate relationships, or rate–equilibrium relationships. An example of the latter is shown in Fig. 2.

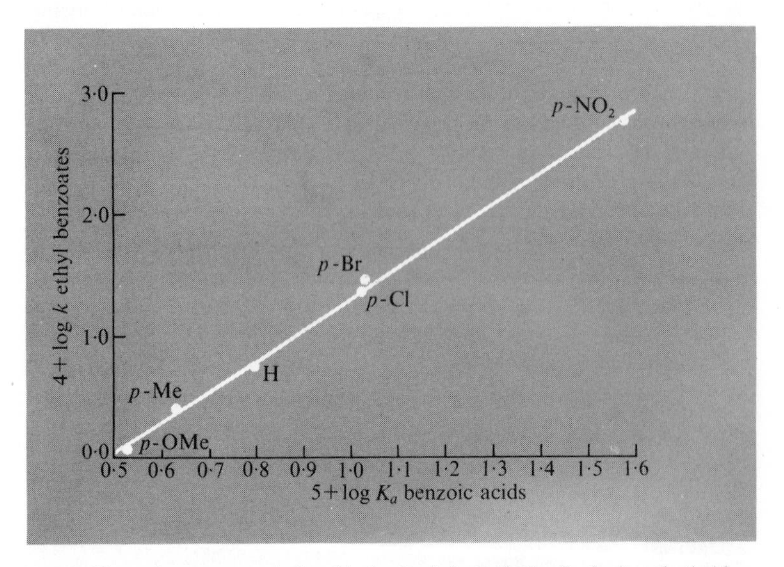

FIG. 2. Linear free-energy relationship for basic hydrolysis of substituted ethyl benzoates and ionization of substituted benzoic acids. k values (l mol^{-1} sec^{-1}, 85 per cent ethanol w/w, 25°C), Ingold and Nathan (1936); K_a values, Dippy (1939).

Such relationships are usually referred to as *linear free-energy relationships*. The appropriateness of this term is easily seen. For equilibria equation (1) applies,

$$\log K = -\frac{\Delta G^{\ominus}}{2{\cdot}303\,RT} \tag{1}$$

where $\Delta G^{\ominus}$ is the standard free-energy change of reaction. For chemical rate processes

$$\log k = \log \frac{RT}{N_A h} - \frac{\Delta G^{\ddagger}}{2{\cdot}303\,RT} \tag{2}$$

where $\Delta G^{\ddagger}$ is the standard free energy of activation. (See p. 7 for explanation of other symbols.) A relationship between logarithms of K or k (at constant temperature) is thus essentially a relationship between free energies. Each set of rate or equilibrium constants is termed a *reaction series*.

These relationships are so numerous and important that the term linear free-energy relationships (LFER) is very often used to cover the whole field of correlation analysis in organic chemistry. However many of the correlation equations are not LFER in the restricted sense of a relationship involving logarithms of rate or equilibrium constants on both sides of the equation. Further, relationships which are by no means linear are often interesting. It is better to use 'correlation analysis' for the whole subject, and to use the term LFER more sparingly.

Correlation analysis in organic chemistry has been widely practised since the early 1930s when Hammett at Columbia and Burkhardt at Manchester discovered linear relationships involving $\log k$ or $\log K$ for a number of systems. This work led to the formulation of the Hammett equation which describes the influence of polar *meta*- or *para*-substituents on the side-chain reactions of benzene derivatives. This equation, and its elaboration and refinement in various ways, is the subject of Chapter 2.

The occurrence of steric as well as polar effects in aliphatic systems and *ortho*-substituted aromatic systems complicates the devising of correlation equations. If we look for simple LFER for these systems we are soon disappointed. A typical example is shown in Fig. 3. There is clearly no simple relationship between the rates of basic hydrolysis of the ethyl esters of the alkanoic acids and the strengths of the acids themselves. Little progress was made until the early 1950s, when Taft made an excellent start in satisfactory correlation analysis in this area. This topic is often referred to as 'the separation of polar, steric, and resonance effects' and is the subject of Chapter 3.

From the work of Hammett and of Taft has come a variety of substituent and reaction parameters of great value in summarizing and understanding the influence of molecular structure on chemical reactivity. The substituent constants have found application in fields very different from those of rates and equilibria of organic reactions. They have been applied extensively in optical spectroscopy (infrared, visible, and ultraviolet), nuclear magnetic resonance spectroscopy (^{1}H, ^{19}F, and other nuclei), and the mass spectrometry of organic compounds. These matters are dealt with in Chapter 4. Further, with appropriate additional considerations, the fields of enzymology and of structure–activity relationships for drugs have been considerably illuminated

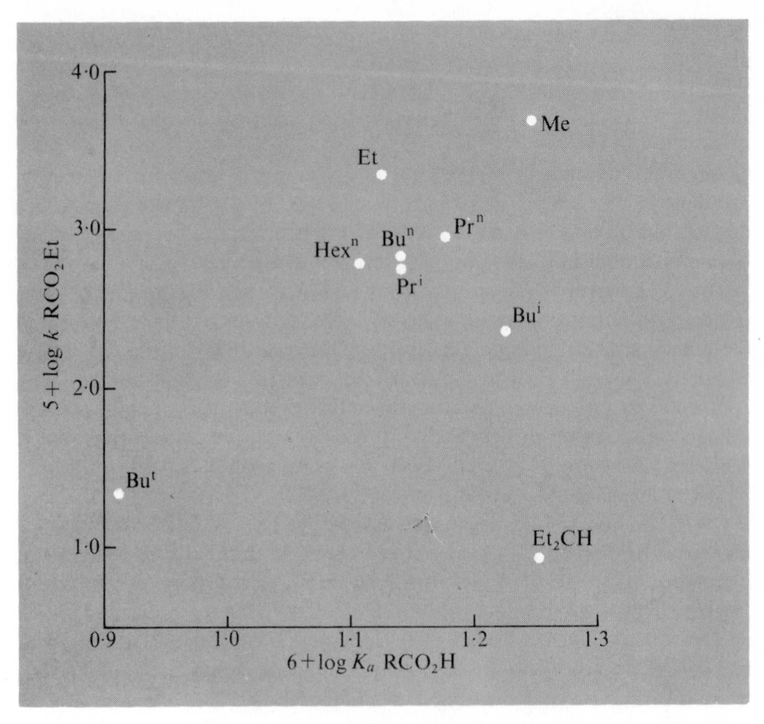

FIG. 3. Lack of linear free-energy relationship for basic hydrolysis of ethyl alkanoates and ionization of alkanoic acids. k values ($l\ mol^{-1}\ sec^{-1}$, 70 per cent acetone v/v, 24·8°C), Davies and Evans (1940); K_a values, Dippy (1939).

by the application of parameters derived from organic chemistry (Chapter 7).

Organic reactions may be regarded as involving an organic *substrate* reacting with a *reagent* in the presence of a *solvent*. The type of correlation analysis described so far is concerned with reaction series in which the structure of the substrate is varied by the introduction of substituents, the solvent and the reagent remaining constant. We can also consider the influence of the solvent or the influence of the reagent on the reactivity of a given substrate. Correlation analysis is also valuable here, although much less highly developed and successful. We consider solvent effects in Chapter 5 and the effect of the reagent in Chapter 6.

Correlation analysis is inevitably much concerned with the calculation of slope and intercept terms in simple linear regression, and the corresponding quantities in multiple regression. Further, estimates of the reliability or success of correlations and the significance of deviations are of great importance. Much misapprehension arises from failure to appreciate matters of elementary statistics. An introduction to the relevant topics in statistics is therefore given as an Appendix. This will help you to appreciate the meaning of widely-used terms, such as correlation coefficient, and indicates the steps necessary to make the statistical side of correlation analysis secure. You are urged to give the Appendix a first reading before starting Chapter 2, and to consult it thereafter from time to time.

Terminology (including symbols and signs) for substituent effects

I am going to assume familiarity with the general ideas used in electronic explanations of the effect of structure upon reactivity. (Consult the general references in the Bibliography if necessary.) However, this is a field in which a variety of terminology is used and there is considerable confusion. We shall try to simplify the situation as far as possible.

The *polar effect* of a substituent comprises all the processes whereby the substituent may modify the electrostatic forces operating at a reaction centre, relative to a standard substituent, which is often, but not always, a hydrogen atom. These forces may be governed by charge separations arising from differences in the electronegativity of atoms (leading to the presence of dipoles), the presence of unipoles, or electron delocalization. Field (or direct), inductive (through-bond polarization involving σ- or π-electrons), and conjugative effects may in principle be distinguished. Because of the difficulty of distinguishing between field and through-bond effects in practice, the term *inductive effect* is often used to cover both, and is normally so used in this book. For conjugative substituent effects the general term *resonance effect* will be used. The term has for long been widely used in connection with LFER, and avoids making the distinction between mesomeric (time-permanent) and electromeric (time-variable) effects, which together constitute the tautomeric or total conjugative effect. 'Resonance' is liable to be misconstrued in terms of oscillation between discrete structures but if this error be avoided, the word is the most useful single term available to us. Thus the prime use of the term resonance effect is to cover conjugation of the substituent with the delocalized system, which (*a*) may, or (*b*) may not, include the functional group.

NO_2— —NH_2 ⟷ ^{-}O $\overset{+}{N}$= =$\overset{+}{N}H_2$ ^{-}O

3

MeO— —CH_2CO_2H ⟷ $Me\overset{+}{O}$= $^{-}$ —CH_2CO_2H

4

Structure **3** is an example of (*a*) and **4** of (*b*). We shall also use the term resonance effect in connection with (*c*), the conjugation of the functional group with the delocalized system, e.g. **5**.

$$C_6H_5{-}CO_2H \longleftrightarrow {}^{+}C_6H_5{=}C(O^-)(OH)$$

5

It will be used in this sense in Chapter 3. The precise meaning of resonance effect may thus only be understood in context.

Hyperconjugation or the *hyperconjugative effect*, as the name implies, is historically and commonly regarded as a special kind of resonance effect involving primarily delocalization of electrons in C—H bonds adjacent to an unsaturated system. The actual nature of this substituent effect, however, must be regarded as still in doubt. The term simply means a special effect of α-hydrogen atoms in the same sense as conjugative electron release. An analogous effect of C—C bonds is also sometimes invoked; this is C—C hyperconjugation, in contrast to C—H hyperconjugation.

Steric effects are caused by the intense repulsive forces operating when two non-bonded atoms approach each other so closely that non-bonded compressions are involved. A *primary steric effect* of a substituent is the direct result of the difference in compression energy which arises because the molecule differs in structure from a suitable standard molecule in the vicinity of the reaction centre. A *secondary steric effect* involves the moderation of a polar effect or resonance effect by non-bonded compressions.

In discussing the influence of any substituent effect one must consider differentially interactions in initial and transition states for rate processes, and in initial and final states for equilibria.

We turn now to the question of symbols and signs for the electronic effects of substituents. Our choice of terms above enables us to select *I* for the inductive effect and *R* for the resonance effect. We shall avoid describing electron delocalization by symbols such as *M*, *E*, *T*, *C*, and *K* which are widely used elsewhere.

The question of sign convention is more difficult. The situation is tangled, as situations regarding sign-convention often are. The Ingold sign convention is widely used in organic chemistry, the world over. Its basis lies in associating *electronegativity* (relative to the hydrogen atom) with a *negative* sign and *electropositivity* with a *positive* sign. Thus the nitro group is described as electron-withdrawing by virtue of its $-I$ and $-M$ effects (to use Ingold's choice of symbol). For correlation analysis this convention is inconvenient for it is in contradiction to the universally accepted sign convention for polar substituent *constants*: an electron-attracting group is given a *positive* substituent constant, and an electron-releasing group is given a *negative* substituent constant. (You will appreciate this better when you have read some of Chapter 2.) We shall avoid the contradiction by adopting the other sign convention for electronic effects. Thus NO_2 will be a $+I$, $+R$ substituent; Cl, a $+I$, $-R$ substituent, etc. You must be careful if you are used to reading about electronic effects in general textbooks of organic chemistry or physical organic chemistry.

Other symbols

We have already mentioned that all chapters are concerned with measures of the success of correlations and the following symbols will be used throughout:

r correlation coefficient in a simple linear regression
R correlation coefficient in a multiple regression
s standard deviation of the estimate of the dependent variable.

R is also used for the gas constant and, as discussed already, for the resonance effect, but there will be no risk of confusion. Other symbols include: T (Kelvin scale temperature), N_A (Avogadro constant), h (Planck's constant), **k** (Boltzmann constant).

2. The Hammett equation

Introduction

IN Chapter 1 we saw how various authors discovered linear relationships between logarithms of rate or equilibrium constants of organic reactions. Many such relationships involve side-chain reactions of *meta*- or *para*-substituted benzene derivatives, and may be summarized by a very simple equation devised by Hammett (1937, 1970). This simple behaviour is due to the influence of these substituents on side-chain reactivity being almost entirely *polar* in nature, cf. *ortho*-substituted compounds and aliphatic systems, in which large steric effects may occur (see Chapter 3).

The Hammett equation takes the forms:

$$\log k = \log k^0 + \rho\sigma \qquad (3)$$

$$\log K = \log K^0 + \rho\sigma \qquad (4)$$

where k or K is the rate or equilibrium constant, respectively, for a side-chain reaction of a *meta*- or *para*-substituted benzene derivative. The symbol k^0 or K^0 denotes the statistical quantity approximating to k or K for the 'unsubstituted' or 'parent' compound (see below). The *substituent constant*, σ, measures the polar effect (relative to hydrogen) of the substituent (in a given position, *meta* or *para*) and is, in principle, independent of the nature of the reaction. The *reaction constant*, ρ, depends on the nature of the reaction (including conditions such as solvent and temperature) and measures the susceptibility of the reaction to polar effects.

The ionization of benzoic acids in water at 25°C is chosen as a standard process, for which ρ is defined as 1·000. The value of σ for a given substituent is $\log(K_a/K_a^0)$, where K_a is the ionization constant of the substituted benzoic acid and K_a^0 is that of benzoic acid itself. It should now be clear that the Hammett equation is a linear free-energy relationship.

To apply the Hammett equation to a given reaction, rate or equilibrium data must be available for the parent compound and several (preferably three or more) *meta*- or *para*-substituted compounds involving substituents of known σ. $\log k$ or $\log K$, as appropriate, is plotted against σ, the parent compound providing a data point at $\sigma = 0$. The best straight line is obtained, preferably by the method of least squares (see Appendix), and values of ρ (the slope) and $\log k^0$ or $\log K^0$ (the intercept at $\sigma = 0$) are calculated. The success of the correlation is commonly assessed (following Jaffé, 1953) by calculating the correlation coefficient, r, and the standard deviation, s, of the estimated values of $\log k$ or $\log K$ (see Appendix). Selected values of σ and ρ are given in Tables 1 and 2. (Figs. 1 and 2 in Chapter 1, pp. 1, 2, are effectively Hammett plots.)

TABLE 1

Selected values† of σ

Substituent	σ_m	σ_p	*Substituent*	σ_m	σ_p
Me	−0·07	−0·17	OMe	0·12	−0·27
Ph	0·06	−0·01	OPh	0·25	−0·32
CF_3	0·43	0·54	SH	0·25	0·15
CN	0·56	0·66	SMe	0·15	0·00
COMe	0·38	0·50	F	0·34	0·06
CO_2H	0·37	0·45	Cl	0·37	0·23
NH_2	−0·16	−0·66	Br	0·39	0·23
NMe_2	(−0·21)	−0·83	I	0·35	0·28
NO_2	0·71	0·78	$\overset{+}{N}Me_3$	0·88	0·82
OH	0·12	−0·37	$\overset{+}{S}Me_2$	1·00	0·90

† Based directly on ionization of substituted benzoic acids (except in the case of σ_m for NMe_2) and mainly from McDaniel and Brown (1958). See also the compilation by Exner (1972). These values should not be used uncritically; see the estimated limits of uncertainty given by McDaniel and Brown.

It is difficult to say what proportion of reactions conform reasonably well to the Hammett equation in the form described above. Jaffé (1953), in a review of the Hammett equation, examined its application to about 400 reaction series. On the basis of correlation coefficients he concluded that about 70 per cent of the correlations were 'satisfactory' or 'excellent' (see Appendix). For reaction series in these categories the Hammett relationship has a mean precision of about 15 per cent in k or K, although in some reaction series certain substituents may show large deviations. About 30 per cent of the correlations, however, were classified by Jaffé as 'fair' or 'poor' and this suggests that many rate or equilibrium data are outside the scope of the Hammett equation in its original form and mode of application.

We shall go on shortly to see how the Hammett equation may be refined, but first it is important to get some appreciation of σ- and ρ-values on the basis of Tables 1 and 2. (Certain points of detail in the determination of some of the ρ-values in Table 2 will be explained later, p. 12 et seq.)

The Hammett constants

The substituent constant, σ

Electron-withdrawing substituents have positive values of σ and electron-releasing substituents have negative values. The σ-scale covers roughly the numerical range 0 ± 1. The interpretation of values in simple electronic terms is often quite easy. For Cl $\sigma_m > \sigma_p$ because the $+I$ effect is opposed by the $-R$ effect for p-Cl.† By contrast $\sigma_m < \sigma_p$ for NO_2, because the $+I$ effect is reinforced by the $+R$ effect for p-NO_2. With OMe the $+I$ effect gives σ_m a

† See p. 5 for discussion of symbols and signs for electronic effects.

TABLE 2

Selected values† of ρ

Reaction	*Solvent and temperature, °C*	*ρ*
Equilibria		
$ArCO_2H \rightleftharpoons ArCO_2^- + H^+$	H_2O, 25	1·00
$ArCH_2{\cdot}CO_2H \rightleftharpoons ArCH_2{\cdot}CO_2^- + H^+$	H_2O, 25	0·49
$ArOH \rightleftharpoons ArO^- + H^+$	H_2O, 25§	2·11
$Ar\overset{+}{N}H_3 \rightleftharpoons ArNH_2 + H^+$	H_2O, 25§	2·94
Rates		
$ArCO_2Et + OH^- \longrightarrow ArCO_2^- + EtOH$	85% EtOH, 25	2·54
$ArCO_2H + MeOH \xrightarrow{H^+} ArCO_2Me + H_2O$	MeOH, 25	−0·52
$ArNH_2 + PhCOCl \longrightarrow ArNH{\cdot}COPh + HCl$	C_6H_6, 25§	−3·21
$ArCOCl + PhNH_2 \longrightarrow ArCO{\cdot}NHPh + HCl$	C_6H_6, 25	1·18
$ArCMe_2Cl \longrightarrow Ar\overset{+}{C}Me_2 + Cl^-$	90% Me_2CO, 25‡	−4·54
$ArCH_3 + \dot{C}Cl_3 \longrightarrow Ar\dot{C}H_2 + CHCl_3$	C_6H_5Cl, 50‡	−1·46
$ArH + Cl_2 \longrightarrow ArCl + HCl$	AcOH, 25‡	−10·0
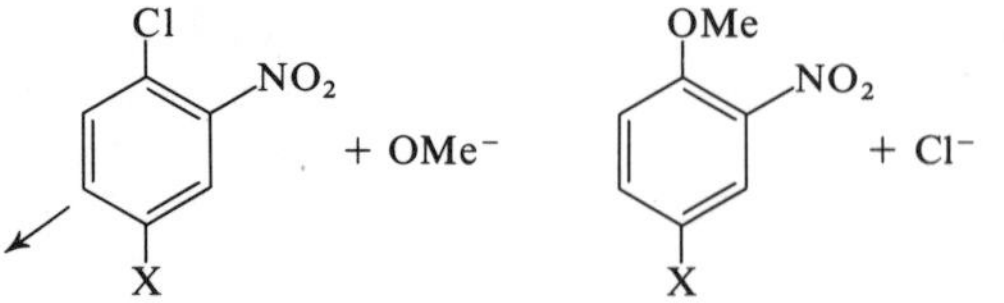 4-X-2-nitrochlorobenzene (Cl, NO_2, X) + OMe^- ⟶ 4-X-2-nitroanisole (OMe, NO_2, X) + Cl^-	MeOH, 50§	3·90

† From various sources. Compilations of ρ-values are given by Jaffé (1953), van Bekkum, Verkade, and Wepster (1959), Wells (1963), and Exner (1972). Values vary slightly, depending on number of compounds considered and method of calculation. A thorough-going compilation should include, for each reaction series, values of *n* (number of compounds), *r* (correlation coefficient), *s* (standard deviation of the estimate), and s_ρ (standard deviation of ρ); see Appendix. Values of $\log k^0$ (or K^0) are also of interest (but not unless their units are clearly stated).

‡ Requires use of σ^+.

§ Requires use of σ^-.

small positive value, but for *p*-OMe the $-R$ effect is so strong that σ_p has a substantial negative value. For NH_2 the situation is somewhat similar except that for *m*-NH_2 the indirect operation of the $-R$ effect through 'relay' is strong enough to make σ_m negative (see structure **6**).

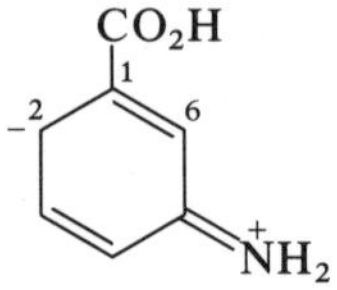

CO_2H at 1; − at 2; 6; $\overset{+}{N}H_2$

6

Influence of − *at* 2 (*and* 6) *relayed to* 1

For CH_3 the σ-values are negative and numerically $\sigma_m < \sigma_p$. This is often ascribed to the operation of $-R$ 'hyperconjugation' from the *para*-position (see structure **7**) but this interpretation is not everywhere accepted.

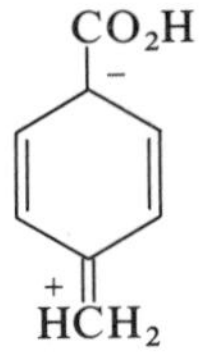

CO_2H

$\overset{+}{H}C H_2$

7

The reaction constant, ρ

A reaction which is facilitated by reducing the electron density at the reaction centre has a positive value of ρ, and one facilitated by increasing the electron density at the reaction centre has a negative value. The ρ-scale covers roughly 0 ± 4. The qualitative interpretation in mechanistic terms of the ρ-values in Table 2 is fairly straightforward. (On p. 27 we shall see how ρ-values may provide valuable evidence of mechanism.)

The release of protons by anilinium ions and the approach of OH^- ions to the carbonyl carbon atom of ethyl benzoates are understandably facilitated by electron withdrawal from the reaction centre. The nucleophilic behaviour of aniline towards benzoyl chloride and the electrophilic behaviour of benzoyl chloride towards aniline (in their reaction to give benzanilide) are naturally affected in opposite ways by electron withdrawal from the reaction centre. The highly negative value of ρ for the effect of substituents in aniline perhaps suggests that a very polar activated complex is formed (see structure **8**), with extensive formation of the N—C bond.

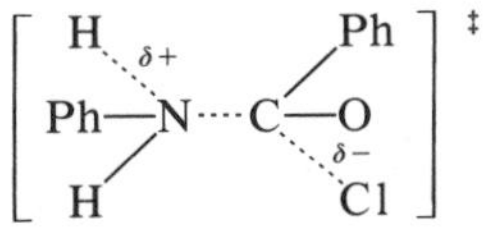

H δ+ Ph

Ph—N····C—O

δ−

H Cl ‡

8

The small ρ-value for the acid-catalysed esterification of benzoic acids in methanol is readily interpreted in terms of the step-wise nature of the reaction mechanism. The essential features of the mechanism are (i) the formation of a small concentration of protonated acid, $PhC(OH)^+OH$, effectively in equilibrium with the bulk of the carboxylic acid, (ii) the rate-limiting nucleophilic attack of methanol on the protonated acid to form the methyl ester.

The observed rate constant is equal to the product of the equilibrium constant for (i) and the rate constant for (ii). The observed ρ-value is the sum of the ρ-values relating to (i) and (ii). Reducing the electron density at the reaction centre will hinder the protonation but facilitate the nucleophilic attack. The two components of the resultant ρ-value are thus of opposite sign. Hence the observed ρ-value is small, even though (by analogy with other reactions) the component ρ-values are probably substantial.

We shall return to this Table shortly.

The duality of substituent constants

We have already mentioned briefly that in some reaction series certain substituents show marked deviations from the Hammett equation. These are commonly *para*-substituents of considerable $-R$ or $+R$ effect. Hammett found that the *p*-NO_2 group showed deviations in the reactions of anilines and phenols. The deviations were systematic in that an exalted σ-value of ca 1·27 seemed to apply, compared with the value of $\sigma = 0{\cdot}78$ based on the ionization of *p*-nitrobenzoic acid. Other examples were soon discovered and it became conventional to treat them similarly in terms of a 'duality of substituent constants'. We must discuss this behaviour in detail and see how it pointed the way to a first refinement of the Hammett equation.

When σ-values based on the ionization of benzoic acid are used, deviations may occur with the $+R$ *para*-substituents for reactions involving $-R$ electron-rich reaction centres and with $-R$ *para*-substituents for reactions involving $+R$ electron-poor reaction centres. The explanation of these deviations is in terms of 'cross-conjugation', that is conjugation involving substituent and reaction centre.

In the ionization of *p*-nitroanilinium ion:

$$p\text{-}NO_2{\cdot}C_6H_4{\cdot}\overset{+}{N}H_3 \rightleftharpoons p\text{-}NO_2{\cdot}C_6H_4{\cdot}NH_2 + H^+ \qquad (5)$$

the free base is stabilized by delocalization of electrons involving the canonical structure **9**. An analogous structure is not possible for the *p*-nitroanilinium

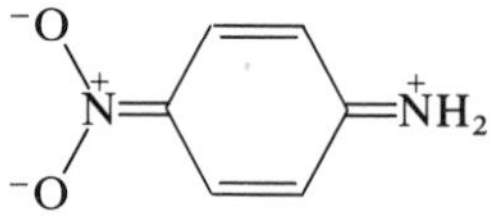

9

ion. In the ionization of *p*-nitrophenol:

$$p\text{-}NO_2{\cdot}C_6H_4{\cdot}OH \rightleftharpoons p\text{-}NO_2{\cdot}C_6H_4O^- + H^+ \qquad (6)$$

analogous delocalization is possible in both species, but is more marked in the *p*-nitrophenate ion. Thus in both the aniline and phenol system *p*-NO_2 is effectively more electron-attracting than in the ionization of *p*-nitrobenzoic acid:

$$p\text{-}NO_2{\cdot}C_6H_4{\cdot}CO_2H \rightleftharpoons p\text{-}NO_2{\cdot}C_6H_4{\cdot}CO_2^- + H^+ \qquad (7)$$

where the reaction centre is incapable of a $-R$ effect, and indeed shows a small $+R$ effect, **10**. Other $+R$ *para*-substituents requiring exalted σ-values when

$O_2\overset{+}{N}$–C_6H_4=$C(O^-)OH$ (quinonoid)

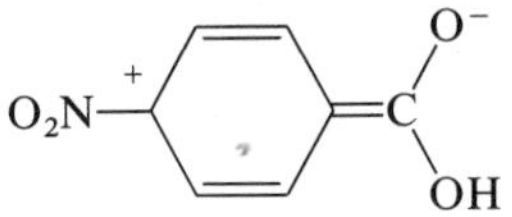

10

in association with $-R$ electron-rich reaction centres include CN, CO_2H, CO_2Me, and $SO_2{\cdot}Me$.

An example of a reaction series in which exalted σ-values are required by $-R$ *para*-substituents is provided by the rate constants for the solvolysis of substituted t-cumyl chlorides, $ArCMe_2Cl$. This reaction follows an S_N1 mechanism, with intermediate formation of the cation $Ar\overset{+}{C}Me_2$. A $-R$ *para*-substituent such as OMe may stabilize the activated complex, which resembles the carbonium–chloride ion pair, through delocalization involving structure **11**. Such delocalization will clearly be more pronounced than in the species

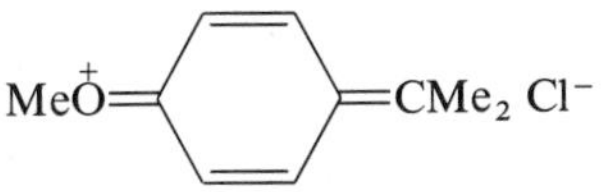

$Me\overset{+}{O}$=C_6H_4=CMe_2 Cl^- (quinonoid)

11

involved in the ionization of *p*-methoxybenzoic acid, which has a reaction centre of feeble $+R$ type. The effective σ-value for *p*-OMe in the solvolysis of t-cumyl chlorides is thus $-0{\cdot}78$, compared with the value of $-0{\cdot}27$ based on the ionization of benzoic acids. Other $-R$ *para*-substituents requiring exalted σ-values when in association with $+R$ electron-poor reaction centres include Me, OH, NH_2, SMe, and Hal.

The special substituent constants for $+R$ substituents are denoted σ^-, and those for $-R$ substituents are denoted σ^+. Values of the former may be based either on the ionization of anilinium ions or of phenols in water. (For some substituents the values based on the two systems are slightly different.) In each case the Hammett line is established with *meta*-substituents and $-R$ *para*-substituents. Values of σ^+ were based by Brown and Okamoto (1958) on the rates of solvolysis of t-cumyl chlorides in 90 per cent acetone–water at 25°C. The Hammett line is established with *meta*-substituents and $+R$ *para*-substituents. (In establishing the Hammett line for σ^-, strong $+R$ substituents are not used in the *meta*-position and for σ^+ strong $-R$ substituents are not used in the *meta*-position, in case there is a relayed effect. For such substituents σ_m^- or σ_m^+ values may subsequently be obtained. Brown calculated σ_m^+ or σ_p^+ for *all* substituents from the Hammett line, but the validity of the small corrections applied to substituents already used to establish the line is doubtful.)

Selected values of σ_p^- and σ_p^+ are shown in Table 3, with ordinary σ_p-values

TABLE 3

Selected values† *of* σ_p, σ_p^+, *and* σ_p^-

Substituent	σ_p	σ_p^+	*Substituent*	σ_p	σ_p^-
Me	−0·17	−0·31	CF_3	0·54	(0·65)
Bu^t	−0·20	−0·26	CN	0·66	0·88
NH_2	−0·66	(−1·3)	COMe	0·50	0·84
OH	−0·37	(−0·92)	$CONH_2$	0·38	0·61
OMe	−0·27	−0·78	CO_2H	0·45	0·73
F	0·06	−0·07	$SiMe_3$	−0·07	0·17
Cl	0·23	0·11	NO_2	0·78	1·24
Br	0·23	0·15	$SO_2{\cdot}Me$	0·72	0·98
I	0·28	0·14	$SO_2{\cdot}NH_2$	0·57	(0·94)

† From the compilation by Exner (1972).
σ_p^+ values based on t-cumyl chloride solvolysis, with exception of NH_2 and OH.
σ_p^- values based on ionization of phenols, with exception of CF_3 and SO_2NH_2 (ionization of anilinium ions).

for comparison. In the main the values require no further comment here. The negative value of σ^+ for *p*-F indicates the importance of the $-R$ effect of this substituent.

Values of σ^- have been applied extensively to the correlation of the reactions of aniline and phenol derivatives, carbanions, and free radicals, and have also been applied to nucleophilic aromatic substitution, i.e. where there is no side-chain, and reaction occurs at a ring carbon (see Table 2).

Values of σ^+ have been applied extensively to solvolyses, aldehyde reactions, basicities of carbonyl compounds and amides, and radical reactions, and most notably to electrophilic aromatic substitution (see Table 2). Such reactions are usually characterized by large negative ρ-values.

The use of σ^+ and σ^- greatly extended the range of applicability of the Hammett equation and often gave information about the nature of the transition state in a particular reaction: the need to use σ^+ or σ^- indicated extensive delocalization. However, certain limitations of σ^+ and σ^- appeared and the inherent artificiality of the 'duality of substituent constants' was recognized.

The variability of cross-conjugation

The attempt to deal with cross-conjugation by selecting σ^+, σ, or σ^- in any given case is artificial. The contribution of the resonance effect of a substituent relative to its inductive effect must in principle vary continuously as the electron-demanding quality of the reaction centre is varied, i.e. whether it is electron-rich or electron-poor. A sliding-scale of substituent constants would be expected for each substituent having a resonance effect and not just a pair of discrete values: σ^+ and σ for $-R$, or σ^- and σ for $+R$ substituents.

How is this idea to be put into practice by modifying the Hammett equation or the way it is used? Care is needed in this matter because clearly the value of the Hammett equation will be reduced if there is a proliferation of different constants characteristic of a given substituent. In the limit σ would become characteristic of the reaction as well as the substituent and the whole system of empirical correlation would then be greatly reduced in value.

Several types of treatment have emerged.

The Yukawa–Tsuno equation

Yukawa and Tsuno (1959) proposed a method for dealing with $-R$ substituents in their influence on reactions which are more electron-demanding than the ionization of benzoic acid. They suggested that values of $(\sigma^+ - \sigma)$ would provide a scale of enhanced resonance effects, and they modified the Hammett equation as follows:

$$\log k = \log k^0 + \rho[\sigma + r(\sigma^+ - \sigma)] \tag{8}$$

(A corresponding equation for equilibria may also be written.) The equation implies the multiple linear correlation of $\log k$ with σ and $(\sigma^+ - \sigma)$. (For $+R$ *para*-substituents and for all *meta*-substituents except those of powerful $-R$ effect, simple linear correlation with σ is effectively involved.) The quantity r (do not confuse this with the correlation coefficient) is a proportionality constant giving the contribution of the enhanced resonance effect for $-R$ substituents. If $r = 0$ the equation reduces to the simple Hammett equation, and if $r = 1$ it corresponds to straightforward correlation with σ^+. Values of $r > 1$ are also possible for reactions which are more electron-demanding than the solvolysis in aqueous acetone of t-cumyl chlorides.

Multiple regression is used to obtain ρ and r (see Appendix). The equation has been applied by Yukawa and Tsuno, and others to numerous reactions, e.g. aromatic substitution, additions to carbon–carbon multiple bonds and carbonyl groups, and is often more successful than the simple Hammett or Brown treatment. The degree of success must be properly assessed by statistical methods (see Appendix).

The quantities ρ and r are in principle independent, although some relationship between them might perhaps be expected. For limited selections of reactions it is possible to discern a tendency for r to increase as ρ becomes more negative. However, for any large range of reactions no clear relationship can be seen. It has been suggested that r is a more intelligible and useful quantity than ρ, and that similarity in r for two reactions indicates a similarity in transition state structure, even though the ρ-values may be very different.

A corresponding equation with σ^- constants to deal with the influence of $+R$ substituents on reactions which are less electron-demanding than the ionization of benzoic acid was formulated by Yoshioka, Hamamoto, and Kubota (1962).

$$\log k = \log k^0 + \rho[\sigma + r(\sigma^- - \sigma)] \quad (9)$$

The applications of this have not been so well explored, but it has been applied to a variety of processes.

Both these equations have since been modified (Yukawa, Tsuno and Sawada, 1966) to use σ^0-values instead of σ-values (see p. 18) but the essential principles are unaltered.

The sliding-scale of sigma values

Van Bekkum, Verkade, and Wepster (1959) strongly criticized the duality of sigma constants as an artificial device to describe 'mesomeric *para* interaction', as they termed cross-conjugation, and they proposed a new method of applying the Hammett equation.

The ionization of benzoic acids is used only to evaluate 'primary' σ-values which can be considered unambiguously 'normal', i.e. free from the effects of cross-conjugation. Only such σ-values are used in calculating ρ for a given reaction series. For other substituents the Hammett line is then used to calculate σ-values relevant to the particular reaction series. The authors applied this method to about eighty reaction series and confirmed the hypothesis of a 'sliding scale' of sigma values arising from cross-conjugation. From the multiplicity of sigma values recorded for each substituent the authors selected a normal, unexalted sigma value, designated σ^n. The assumption was made that $-R$ reaction centres give normal sigma values for $-R$ substituents and $+R$ centres for $+R$ substituents, a small spread of calculated σ^n values being averaged for each substituent.

Eight primary σ-values of *meta*-substituents (including H, $\sigma = 0{\cdot}00$) are of general applicability and *p*-COMe and *p*-NO_2 are considered acceptable for systems in which cross-conjugation can be ruled out (see Table 4). All other *para*-substituents are unacceptable, along with *meta*-substituents such as OH and OMe, whose strong $-R$ effects may cause anomalies even from the *meta*-position. The authors claimed that these restrictions led to striking improvements in the success of Hammett correlations, but this conclusion was

TABLE 4

Selected values† of σ, σ(primary), σ^n, and σ^0

Substituent	σ	σ (*primary*)	σ^n	σ^0
m-Me	−0·07	−0·07	—	−0·07
m-COMe	0·38	0·38	—	0·34
m-NH_2	−0·16	—	−0·04	−0·14
m-NO_2	0·71	0·71	—	0·70
m-OH	0·12	—	0·10	0·04‡
m-OMe	0·12	—	0·08	0·06‡
m-F	0·34	0·34	—	0·35
m-Cl	0·37	0·37	—	0·37
m-Br	0·39	0·39	—	0·38
m-I	0·35	0·35	—	0·35
p-Me	−0·17	—	−0·13	−0·15
p-COMe	0·50	0·50§	—	0·46¶
p-NH_2	−0·66	—	−0·17	−0·38
p-NO_2	0·78	0·78§	—	0·82¶
p-OH	−0·37	—	−0·18	−0·13‡
p-OMe	−0·27	—	−0·11	−0·16‡
p-F	0·06	—	0·06‖	0·17
p-Cl	0·23	—	0·24	0·27
p-Br	0·23	—	0·27	0·26
p-I	0·28	—	0·30	0·27

† For sources of σ values see Table 1. Values of σ (primary) and σ^n based on van Bekkum, Verkade, and Wepster (1959). Values of σ^0 from Taft (1960).

‡ For reactions in non-hydroxylic solvents.

§ Used for reactions where cross-conjugation does not occur.

¶ For reactions in water and in aqueous organic mixtures.

‖ Regarded as suspect by van Bekkum, Verkade, and Wepster (1959); they considered the proper value should be ca. 0·17.

not entirely secure because insufficient attention was paid to the increase in correlation coefficient due to reducing the degrees of freedom of the regression (see Appendix). There is, however, no doubt that the authors' main chemical ideas are sound.

Values of σ^n are in Table 4. Note the large differences between σ^n- and σ-values based on benzoic acid ionization for *p*-NH_2, *p*-OH, and *p*-OMe, i.e. the σ-values are much affected by cross-conjugation involving CO_2H.

The work of van Bekkum, Verkade, and Wepster is important but it tends towards the situation in which the 'substituent constant' becomes reaction dependent and correlation analysis is devalued.

The separation of inductive and resonance effects

Hammett σ-values measure the resultant of inductive and resonance effects. Taft and Lewis (1958, 1959) suggested that the resultant effects should be quantitatively separable into inductive and resonance contributions through the following equations.

$$\sigma_m = \sigma_I + \alpha\sigma_R \tag{10}$$

$$\sigma_p = \sigma_I + \sigma_R \tag{11}$$

The inductive effect, given by σ_I, is assumed to operate equally from the *meta*- and the *para*-position. The resonance effect, given by σ_R, contributes to σ_m indirectly, α being the 'relay coefficient'.

Taft and Lewis set up a σ_I scale based on alicyclic and aliphatic reactivities (see Chapter 3). For ordinary Hammett σ-values, based on the ionization of benzoic acid, a value for α of 0·33 was suggested. The authors agreed with van Bekkum *et al.*, however, in regarding the ionization of benzoic acids as not an entirely satisfactory standard process. They proposed the use of 'insulated' reaction series, i.e. reactions in which the reaction centre is incapable of conjugation with the ring at any stage. The ionization of phenylacetic acids is such a series. σ-Values based on such processes were designated σ^0. Selected values are shown in Table 4. The values correspond fairly closely in most cases to the primary σ-values or σ^n-values (van Bekkum *et al.*) as the case may be, but sometimes there are considerable differences, e.g. for *p*-NH_2. The analysis into inductive and resonance effects may also be performed with σ^0-constants, giving σ_R^0-values ($\alpha = 0{\cdot}5$) or with σ^+ and σ^- constants giving σ_R^+ and σ_R^- respectively. Selected values of σ_I, σ_R^0, σ_R^+, and σ_R^- are shown in Table 5.

TABLE 5

Selected values† of σ_I, σ_R^0, σ_R^+, σ_R^-; also σ_I(BA) and σ_R(BA)

Substituent	σ_I	σ_R^0	σ_R^+	σ_R^-	σ_I(BA)	σ_R(BA)
Me	−0·05	−0·10	−0·25	—	−0·06	−0·07
COMe	0·28	0·18	—	0·55‡	0·34	0·05
NH_2	0·12	−0·48	−1·4§	—	0·11	−0·78
NO_2	0·63	0·19	—	0·6‡	0·70	0
OH	0·25	−0·40	−1·2§	—	0·28	−0·68
OMe	0·26	−0·41	−1·02	—	0·31	−0·63
F	0·52	−0·35	−0·57	—	0·56	−0·59
Cl	0·47	−0·20	−0·36	—	0·51	−0·35
Br	0·45	−0·19	−0·30	—	0·50	−0·34
I	0·40	−0·12	−0·25	—	0·43	−0·23

† Based on compilations by Wells (1963), Ehrenson, Brownlee, and Taft (1972), and Exner (1972). These quantities should not be used uncritically; the values proposed by different authors vary appreciably in some cases. For this Table values as far as possible in accord with the data in Tables 1, 3, and 4 have been selected (see especially footnotes ‡ and § below). Note that σ_I(BA) and σ_R(BA) are based on Exner's equations and must not be compared with the others based on Taft's equations.

‡ Values calculated to agree with σ_p^- in Table 3.

§ Values calculated to agree with σ_p^+ in Table 3.

Exner (1966) has carried out a modified analysis of this type, based on the following equations.

$$\sigma_m = \sigma_I + 0{\cdot}33\,\sigma_R \qquad (12)$$

$$\sigma_p = \lambda\sigma_I + \sigma_R \qquad (13)$$

The coefficient λ expresses any difference in the operation of the inductive effect as between the *meta*- and the *para*-position, cf. Taft and Lewis, who assumed essentially $\lambda = 1{\cdot}00$. Exner argued cogently in favour of $\lambda = 1{\cdot}14$, i.e. the inductive effect operates more powerfully from the *para*-position. Exner has tabulated σ_I and σ_R values consistent with σ-values based on the ionization of benzoic acid [see σ_I(BA) and σ_R(BA) in Table 5]. Note that Exner gives $\sigma_R = 0$ for NO_2, i.e. the ordinary conjugation of NO_2 with the ring is very small.

The importance of the separation of σ-parameters into σ_I- and σ_R-type contributions is that it suggests the possibility of a 'dual parameter' treatment for reaction series in terms of the equation:

$$\log(k/k^0) = \rho_I\sigma_I + \rho_R\sigma_R \qquad (14)$$

Many correlations of this type have been reported, by using, for example, σ_R^0. This approach seems ideal, provided that σ_R-type values for the various substituents do not change relatively to each other when the electron-demand of the reaction changes. Unfortunately, comparison of the different scales of resonance parameters finds only limited correlation between them. Ehrenson, Brownlee, and Taft (1972) have suggested that σ_R (from benzoic acid ionization), σ_R^0, σ_R^+, or σ_R^- may be required.

Deviations from the Hammett equation in its various forms

Some reaction series show large deviations with even the most refined modes of applying the Hammett equation. The most general cause of this is complexity of reaction mechanism. If the observed rate constant is effectively an overall rate constant arising from two or more competing mechanisms or is really a complex function of rate constants for a series of reaction steps, deviations are liable to occur. When the substituent is varied over a wide range of polarity, the balance between the various processes is altered. Constancy of mechanism throughout the reaction series is a basic requirement for conformity to the Hammett equation in any of its forms. Analogous ideas apply to equilibrium correlations: the essential structures of reactant and product must be constant throughout the reaction series. We shall now see some examples of major deviations due to complexity of mechanism.

The solvolyses of benzyl chlorides ($ArCH_2Cl$) or tosylates (p-$MeC_6H_4 \cdot SO_2 \cdot OCH_2Ar$) in aqueous organic solvents give poor correlations by the procedure of van Bekkum *et al.*, and calculated σ-values for $-R$ *para*-substituents are absurd, e.g. $\sigma = -1{\cdot}88$ for p-OMe (cf. $\sigma^+ = -0{\cdot}78$). This behaviour is due to the mechanism not remaining constant throughout each reaction series. Probably the parent compounds react mainly by a bimolecular mechanism, with a process involving an intermediate carbonium ion, $Ar\overset{+}{C}H_2$, playing only a small part. Substituents alter the relative importance of the mechanisms, leading to poor correlations even with primary σ-values, and to the apparently tremendous electron-releasing power of $-R$ *para*-substituents, which can stabilize $Ar\overset{+}{C}H_2$ by cross-conjugation, so that reaction occurs almost entirely via the carbonium ion.

Competing mechanisms can lead to a Hammett plot which is concave upwards and shows a minimum, e.g. the denitrosation of substituted *N*-methyl-*N*-nitrosoanilines shown in Fig. 4.

$$ArN(NO)CH_3 + 2H^+ \longrightarrow Ar\overset{+}{N}H_2{\cdot}CH_3 + NO^+ \quad (15)$$

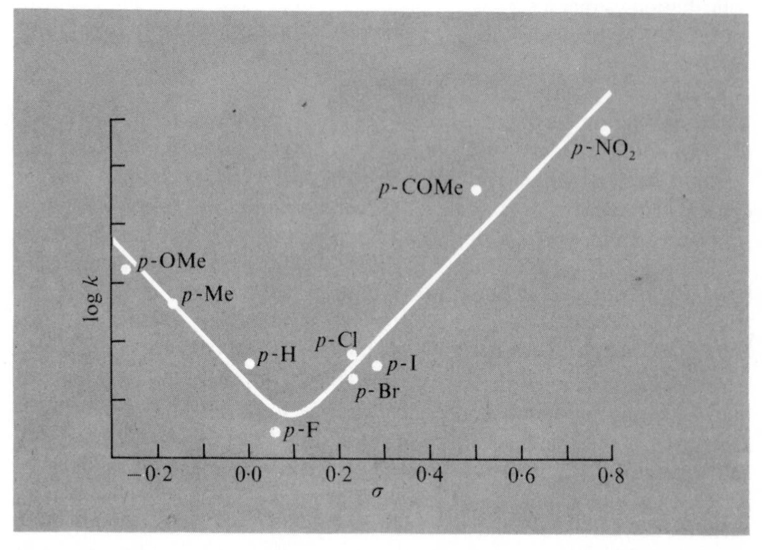

FIG. 4. Hammett plot for denitrosation of substituted *N*-methyl-*N*-nitrosoanilines. After Exner (1972). The divisions of the ordinate are 0·5 unit of $\log k$ apart.

Hammett plots can also be concave downwards and show a maximum. This may happen for a reaction whose mechanism involves a reversible step followed by an irreversible one giving the reaction products. If the observed rate constant is effectively the product of the equilibrium constant for the first step and the rate constant for the second step, then the Hammett equation is still obeyed. This is the case for acid-catalysed esterification of benzoic acids in methanol (as we saw on p. 11), with quite a wide range of substituents.

However, if the scheme:

$$A \underset{k_{-1}}{\overset{k_1}{\rightleftharpoons}} B \tag{16}$$

$$B \xrightarrow{k_2} \text{products} \tag{17}$$

applies, the stationary-state treatment gives

$$k_{obs} = k_1 k_2/(k_{-1} + k_2). \tag{18}$$

Substituents may alter the relative magnitudes of the various constants so that at one extreme $k_{obs} = k_1$, and at the other, $k_{obs} = Kk_2$. The Hammett equation will then break down for k_{obs}. This is the situation for the reaction shown in Fig. 5, the acidic hydrolysis of substituted salicylideneanilines.

$$X\text{-}C_6H_4\text{-}N{=}CH\text{-}C_6H_4(OH) + H_3\overset{+}{O} \longrightarrow X\text{-}C_6H_4\text{-}\overset{+}{N}H_3 + C_6H_4(OH)CHO \tag{19}$$

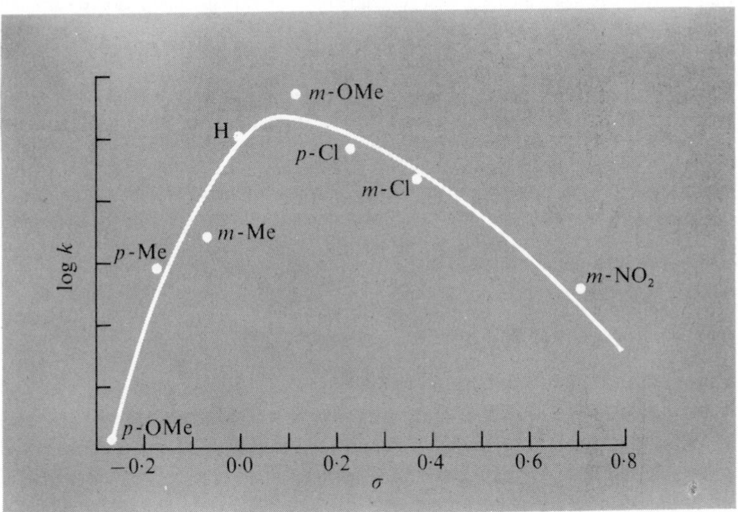

FIG. 5. Hammett plot for acidic hydrolysis of substituted salicylideneanilines. After Exner (1972). The divisions on the ordinate are 0·1 unit of log k apart.

Complexity of mechanism is not the only cause of deviations. Types of deviation and possible causes are discussed by Exner (1972).

The general validity of the Hammett equation

In the previous section our object was to account for deviations by examining features of the deviating systems which distinguish them from those which conform well to the Hammett equation. There is a much wider question: why does the Hammett equation work so well for many systems?

The validity of the Hammett equation and of any other linear free-energy relationship involves subtle matters which have been discussed fully by Wells (1963) and by Ritchie and Sager (1964). Free-energy changes are related to enthalpy and entropy changes by equation (20).

$$\Delta G^{\ominus} = \Delta H^{\ominus} - T\Delta S^{\ominus} \tag{20}$$

In a given reaction series substituents may affect both the enthalpy and the entropy term which contribute to free energy of activation or reaction, i.e. the effect of substituents is given by

$$\Delta\Delta G^{\ominus} = \Delta\Delta H^{\ominus} - T\Delta\Delta S^{\ominus} \tag{21}$$

A linear relationship between the $\Delta G^{\ominus}$ values (i.e. $\log k$ or $\log K$) for two reaction series would be expected to hold only if each series shows one of the following types of behaviour: (*a*) $\Delta S^{\ominus}$ is constant, (*b*) $\Delta H^{\ominus}$ is constant, or (*c*) $\Delta H^{\ominus}$ is linearly related to $\Delta S^{\ominus}$.

For a long time it was thought that even the standard process, the ionization of benzoic acids, did not fall into one of these categories. Recently, however, Bolton, Fleming, and Hall (1972) have claimed that very precise measurements of enthalpies and entropies of ionization suggest conformity to (*c*). Be that as it may, it seems certain that many processes which are well correlated by the Hammett equation do not fall into one of the above categories. The empirical success of the Hammett equation is thus to some extent a mystery. Attempts to understand it involve the statistical mechanics and thermodynamics of chemical equilibria and kinetics. Theories of the influence of polar substituents on reactivity concern the potential energies of reacting systems. Actual substituents, however, are liable to change both kinetic and potential energies (see Taft, 1956). It has been suggested that while such changes may be manifested in both enthalpy and entropy terms, a considerable cancelling of kinetic energy effects occurs when these are combined to give a free-energy term, i.e. changes in $\log k$ or $\log K$ are a good measure of potential energy effects (see Ritchie and Sager, 1964; also Hepler, 1963). While this has not been demonstrated in a rigid and general way, it provides some explanation for the success of the Hammett equation. It is interesting that attempts to develop correlation equations in terms of enthalpies of activation and reaction, or the corresponding entropies, have not been particularly successful.

Further discussion of substituent constants

We have already seen that Hammett substituent constants are well understood qualitatively in terms of the combined influence of the inductive and resonance effects. As mentioned in Chapter 1, 'inductive effect' is a useful umbrella term to cover both field and through-bond polarization effects, the two ways in which we can envisage the influence of a substituent dipole or unipole 'reaching' the reaction centre. Actually thinking in terms of these routes may be somewhat misleading. The electrostatic forces operating at the reaction centre are no doubt governed by the complete electron distribution of the molecule and it is this that is modified by the introduction of a substituent. 'The complete electron distribution of the molecule' is something we cannot deal with. We must therefore simplify the situation in terms of models. The kinds of model which are most obvious are the field-effect model and the through-bond polarization model. These are sometimes discussed as if one of them is 'true' and the other 'false'. This is an erroneous attitude. Both models are gross simplifications: we are only entitled to discuss which of them 'works' in a particular situation.

Organic chemists often use the through-bond polarization model exclusively, and in particular they envisage the polarization of the π-electron system of benzene as playing the most important part. However the transmission of polar effects through saturated rings such as cyclohexane (**12**) or bicyclo [2,2,2]octane (**13**) is comparable with transmission through benzene, and this

CH_2—CH_2
CH_2 CH_2
CH_2—CH_2

12

CH_2—CH_2
CH_2—CH_2
HC CH
CH_2—CH_2

13

suggests either that σ-bonds should be included in the through-bond polarization model or that the field-effect model may have considerable merit, in particular that of simplicity. There is now plenty of evidence that the field-effect model often works quite well.

The field-effect model

Kirkwood and Westheimer (1938) postulated that the field effect from a substituent to a reaction centre was transmitted via the molecular cavity and the surrounding medium. Simplifying assumptions about the shape of the cavity permitted an effective dielectric constant, D_{eff}, to be calculated. The field effect, F (in $\log K_a$ units) for the influence of a unipolar or a dipolar substituent on the ionization of a carboxylic acid is given by the following equations, where Ze is the charge of a unipolar substituent,

$$\mathrm{F} = \frac{e^2 Z}{2{\cdot}303\,\mathbf{k}TrD_{eff}} \tag{22}$$

$$\mathrm{F} = \frac{e\mu\cos\theta}{2{\cdot}303\,\mathbf{k}Tr^2 D_{eff}} \tag{23}$$

μ the dipole moment of a dipolar substituent, r the distance between the substituent and the reaction centre (the location of the carboxyl proton), and θ the angle between the axis of the dipole and the line joining the centre of the dipole and the reaction centre; e is the electronic charge.

The application of the treatment to *para*-substituted benzoic acids (Kirkwood and

Westheimer, 1938) and especially the application to *meta*-substituted benzoic acids (Sarmousakis, 1944), in which the resonance effect was less important, gave encouraging results. The application to saturated ring systems, in which there is no resonance effect, clearly demonstrates the value of the field-effect model. In this connection 4-substituted bicyclo[2,2,2]octanecarboxylic acids are of particular interest as providing a geometrical model for *para*-substituted benzoic acids, without the complication of the resonance effect, cf. **14** and **15**.

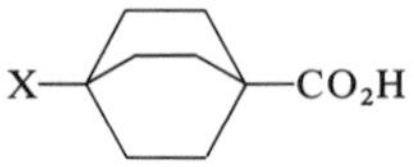

X—[bicyclo[2,2,2]octane]—CO_2H X—[benzene]—CO_2H

14 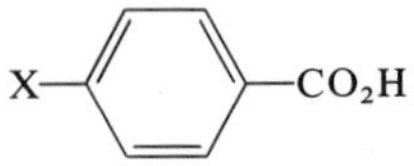 **15**

The transmission of polar effects through saturated ring systems has also been treated in terms of successive polarization of σ-bonds, so that the effect is reduced by a constant factor ε, the attenuation factor. On the whole this treatment accounts less well for the experimental observations.

However, the demonstration of the value of the field-effect model raises certain awkward questions regarding the Hammett equation and its associated parameters. For instance, Taft and Lewis's assumed equality of the inductive effect from the *meta*- and the *para*-position does not accord with the field-effect model, still less Exner's $\lambda = 1{\cdot}14$. Both these seem to require a special polarization of the π-electrons, operating from the *para*-position.

The most awkward question concerns the validity of the Hammett equation, because the quantities r, θ, and D_{eff} of the field model do not change in the same way for all substituents as the reaction is varied. This situation contrasts with the through-bond effect model which is fully compatible with the Hammett equation. Dewar and Grisdale (1962) devised a treatment which eliminated the directional aspect (θ) of the field effect and included the resonance and π-bond polarization effects.

Dewar and Grisdale treatment

The basic equation of this treatment is as follows

$$\sigma_{ij} = F/r_{ij} + Mq_{ij} \tag{24}$$

The σ-value for the influence on a side-chain at position j of a substituent located at position i of an aromatic system is analysed into a field effect term F/r_{ij} and a term Mq_{ij} giving the resonance (mesomeric) and π-bond polarization effects. The field effect is regarded as the effect at ring carbon atom j of the charge induced at carbon atom i by the attached substituent. The distance between the atoms, r_{ij}, is expressed in terms of the carbon–carbon bond length in benzene, assumed to apply to any aromatic system of interest. The quantity q_{ij} is the formal charge at position j produced by attaching CH_2^- at position i; this may be calculated by a simple quantum mechanical method. The field parameter F and the resonance parameter M were calculated for each substituent from the ordinary Hammett σ-values by applying the above equation in the following forms:

$$\sigma_m = \sigma_{31} = F/\sqrt{3} \tag{25}$$

$$\sigma_p = \sigma_{41} = F/2 + 0{\cdot}143M \tag{26}$$

The F and M values for a variety of substituents were then used to predict σ-values for the influence at the 1 position of naphthalene of substituents in the 5, 6, or 7 position and for the influence at the 4 position of biphenyl of substituents in the 3′ or 4′ position.

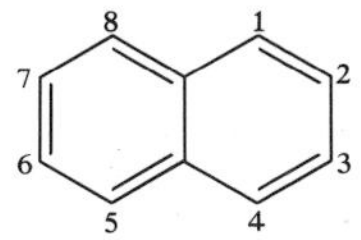

16 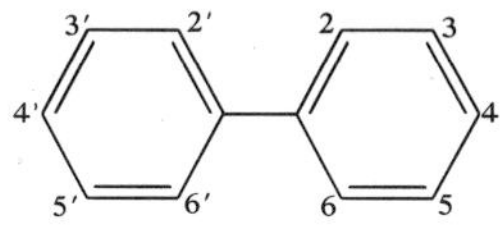 **17**

Predicted σ-values corresponded fairly well with observed reactivities. The treatment has also been applied with some success to naphthalene with the reactive side-chain at the 2 position. The treatment fails badly when extended to a system involving an organic side-chain conjugated with a benzene ring, i.e.

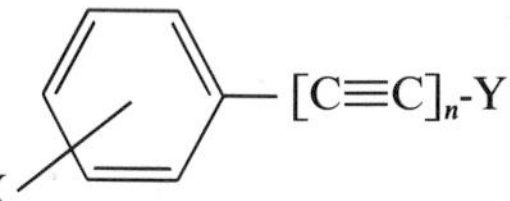

m- or *p*-X [C≡C]$_n$-Y

18

with $n = 1, 2,$ or 3.

Dewar and Grisdale attempted to justify their simplification of the field effect, but it is artificial to neglect angular dependence for a dipolar substituent. The limited empirical success of the treatment can be taken to confirm the importance of the field-effect model.

Further discussion of reaction constants

Dependence on the side-chain

For a given type of reaction , ρ depends strongly on the nature and length of the side-chain. The reactions most commonly used in studying this effect are ionizations of carboxylic acids, their reactions with diazodiphenylmethane (27) and the basic hydrolysis of their esters. In such reactions the introduction

$$ArCO_2H + Ph_2C{:}N_2 \longrightarrow ArCO_2CHPh_2 + N_2 \tag{27}$$

of each CH_2-group decreases ρ by a factor of 0·4–0·5, and the effect of introducing several other atoms or groups into the side-chain differs little from this, e.g. O, S, or SO_2 (Table 6).

TABLE 6

Dependence of ρ on side-chain, temperature, and solvent for reactions of carboxylic acids and esters†

Substrate	*Solvent*	*Temp.* (°C)	ρ
Ionization of acid			
$ArCO_2H$	H_2O	25	1·00
$ArCH_2{\cdot}CO_2H$	H_2O	25	0·49
$Ar[CH_2]_2CO_2H$	H_2O	25	0·21
ArCH:CHCO$_2$H(*trans*)	H_2O	25	0·47
$ArSCH_2{\cdot}CO_2H$	H_2O	25	0·30
$ArSO_2{\cdot}CH_2{\cdot}CO_2H$	H_2O	25	0·25
Reaction of acid with diazodiphenylmethane			
$ArCO_2H$	MeOH‡	30	0·88
$ArCO_2H$	EtOH‡	30	0·94
$ArCO_2H$	Pr^{i}OH‡	30	1·07
$ArCO_2H$	Bu^{t}OH‡	30	1·28
$ArCH_2{\cdot}CO_2H$	EtOH	30	0·40
$Ar[CH_2]_2CO_2H$	EtOH	30	0·22
ArCH:CHCO$_2$H(*trans*)	EtOH	30	0·42
ArC⋮C·CO$_2$H	EtOH	30	0·31
p-ArC$_6$H$_4$·CO$_2$H	EtOH	30	0·22
$ArOCH_2{\cdot}CO_2H$	EtOH	30	0·25
Basic hydrolysis of ester			
$ArCO_2Et$	60% Me_2CO	0	2·66
$ArCO_2Et$	60% Me_2CO	15	2·53
$ArCO_2Et$	60% Me_2CO	25	2·47
$ArCO_2Et$	60% Me_2CO	40	2·38

† For sources of ρ-values see footnote to Table 2. Data for diazodiphenylmethane reaction from Bowden, Chapman, and Shorter (1964), and Buckley, Chapman, Dack, Shorter, and Wall (1968).

‡ Dielectric constants (25°C) are as follows: MeOH, 32·7; EtOH, 24·3; Pr^{i}OH, 18·3; Bu^{t}OH, 12·2.

Conjugated side-chains, however, transmit substituent effects better than saturated chains. A CH:CH-group is roughly equivalent to a CH$_2$-group. Transmission through C⋮C seems to be weaker than through CH:CH. The insertion of a phenylene group decreases ρ by a factor of about 0·25; presumably the effect of a large distance between the reaction centre and the benzene ring bearing the substituent is partially overcome by the extensive conjugation.

Dependence on temperature

For many reaction series the spread of the rate constants becomes smaller as the temperature is increased. Values of σ are defined as temperature-independent. Hence ρ decreases with increase in temperature. It is easy to show that in simple cases ρ should vary as $1/T$. This has been found for various reaction series, although the temperature effect is often rather small (see Table 6). A reversed dependence, i.e. ρ increasing with temperature, is sometimes found but the reactions are usually of complex mechanism.

Dependence on the solvent

Solvent effects on rate and equilibrium constants will be considered in detail in Chapter 5, but the effect on ρ is appropriately discussed briefly here. The effect of solvent on ρ is often very considerable, but is not well understood. The trend is for ρ to decrease as the dielectric constant of the solvent is increased (see Table 6). This corresponds to the reduced importance of electrostatic forces and hence of polar substituent effects in media of high dielectric constant. A simple model predicts a linear relationship of ρ with $1/D$. This is often found to hold approximately, but particular solvents may deviate from the rule.

For the solvent effect the *reactivity–selectivity principle* predicts a relationship between $\log k^0$ and ρ: greater reactivity ($\log k^0$) should be accompanied by lower selectivity (ρ). This is often found to hold approximately for a given reaction, but again there may be individual deviating solvents.

It should be mentioned that variation of σ-values with solvent has been detected for certain substituents, e.g. OH and OMe, see entries and footnotes in Table 4. There has been little systematic study of this, and many of the data do not permit reaction-dependence and solvent-dependence of σ to be clearly distinguished. It seems likely that hydrogen-bonding interactions between substituent and solvent are important.

Applications of the Hammett equation

An obvious application of the Hammett equation is to the prediction of unknown rate and equilibrium constants. In practice relatively little use seems to have been made of this, the Hammett equation being more important as a means of summarizing reactivity data.

The most common application of the Hammett equation is in connection with mechanistic studies. Many authors have sought to show that the influence of *meta*- or *para*-substituents in a given reaction of an aromatic system supports a postulated mechanism or at least does not disprove it. Relevant evidence may be obtained variously from the ρ-value, from the kind of σ-values needed in the correlation, or from the linearity or non-linearity of a particular Hammett plot.

The sign of ρ may be very revealing. Thus the positive ρ-value, 1·57, for the ethanolysis of substituted benzoyl chlorides indicates a bimolecular rather than a unimolecular mechanism which would be assisted by electron-releasing groups. The numerical value of ρ can give useful information about the site of reaction and the nature of the transition state. For instance the high ρ-value, 2·15, for the dissociation of phenylboronic acids, $ArB(OH)_2$, in 25% ethanol indicates that the process involved is

$$ArB(OH)_2 + H_2O \rightleftharpoons ArB(OH)_3^- + H^+ \tag{28}$$

rather than

$$ArB(OH)_2 \rightleftharpoons ArB(OH)O^- + H^+ \tag{29}$$

for which the ρ-value would be comparable with an estimated ρ-value of ca. 1·3 for benzoic acids in the same solvent. The high value of ρ, ~2·1 for the elimination reactions of substituted 2-phenylethyl bromides with alkoxide ions shows that the transition state has much carbanion character, see **19**.

$$\begin{array}{c} \delta_1^- OR \\ \vdots \\ H \quad \delta_2^- \\ \vdots \\ Ar—CH—CH_2Br \end{array}$$

19

The need to use σ^+ or σ^- in correlations often provides valuable evidence (see p. 14). Conclusions are sometimes possible from the fact that σ^+ or σ^- need not be used. (The Yukawa–Tsuno equation presents an alternative approach.)

In discussing deviations from the Hammett equation we have already indicated how non-linearity may serve to detect competitive and consecutive processes.

The last application we consider is the use of σ-values to characterize the electronic behaviour of substituents. Thus the values of σ_m, σ_p, σ_p^+, and σ_p^- of a given substituent reveal much about its inductive and resonance effects. For instance σ_m for $SiMe_3$ is $-0·04$ and σ_p is $-0·07$, but σ_p^- is $+0·17$. This suggests extensive π–d conjugation of the silicon with the ring when the reaction centre is electron-rich. There is similar evidence for groups containing other second-row elements which may be interpreted in the same way. For CF_3, σ_m is 0·43, σ_p is 0·54, and σ_p^- is 0·65; this indicates delocalization involving the ring—CF_3 bond, which suggests C—F hyperconjugation, as in (**20**).

$$+ \; \text{(cyclohexadienylidene)}{=}CF_2F^-$$

20

The extension of the Hammett equation: polysubstituted, polycyclic aromatic, and heterocyclic systems

This chapter has so far dealt with the influence of a single *meta*- or *para*-substituent on the reactivity of a side-chain in a benzene derivative. The effect of polysubstitution, i.e. the reactions of 3,4- or 3,5-disubstituted compounds may be treated either by defining special σ-values for polysubstitution or by examining the application of ordinary σ-values (the problem of 'additivity'). The application of σ-values to 2,3-, 2,4-, or 2,5-disubstituted compounds with a constant *ortho*-substituent may also be examined.

More complicated aromatic systems such as naphthalene, biphenyl, and anthracene may be treated by defining σ-values appropriately for each relative disposition of substituent and reactive side-chain. (Some indication of the role of such studies in connection with the Dewar and Grisdale treatment was given on p. 25.) Often, however, paucity of data prevents the adoption at present of such an approach, but the relevance of ordinary σ-values defined for benzene derivatives can be examined in suitable cases, e.g. the 3 and the 4 position for a reaction centre at position 1 in naphthalene. The same approaches can also be adopted for heterocyclic compounds, e.g. for a reactive side-chain at the 2 position of furan, thiophen, or pyrrole, the effect of substituents at the 4 position may either be treated in terms of special σ-values, or the application of ordinary σ_m-values may be examined.

For all such matters, more extended accounts, e.g. that of Exner (1972), should be consulted.

PROBLEMS

2.1. Carry out a Hammett correlation for the rate constants ($l\,mol^{-1}\,min^{-1}$) of the reaction of substituted benzoic acids with diazodiphenylmethane in ethanol at 30°C. [Data from Benkeser, R. A., DeBoer, C. E., Robinson, R. E., and Sauve, D. M. (1956). *J. Am. chem. Soc.* **78**, 682.]

Substituent	*k*	*Substituent*	*k*
m-Me	0·96	*p*-Br	1·99
H	1·07	*m*-Br	2·60
m-OMe	1·29	*m*-NO_2	5·21
p-Cl	1·67	*p*-NO_2	5·32

Use the σ-values in Table 1. Calculate ρ, $\log k^0$, r, and s by the method of least squares (see Appendix) and also draw the graph. The most educational way of doing the work would be by using an electronic desk calculator.

2.2. (*a*) Use the σ_I- and σ_R-type parameters (Table 5) to interpret in detail the patterns of σ- and σ^0-values (Table 4) and of σ_p^+ values (Table 3) shown by the four halogen substituents in the *meta*- and the *para*-position.
(*b*) Interpret the σ-values for OMe and SMe (Table 1) in terms of the *I* and the *R* effect. Find other LFER data for these groups in the Tables or in the compilations to which reference is made, and interpret these also.
(*c*) Few substituents are characterized by having both distinctive σ_p^+ and σ_p^- values; no groups of this type are given in Table 3. What features must such a group possess? Try to find relevant LFER data and interpret them.

2.3. Plot σ_p against σ_m for the following substituents from Table 1: CF_3, CN, COMe, CO_2H, and NO_2. Use also the data in the appended Table (from the compilation by Exner 1972). Draw the best straight line, and measure its slope.

Substituent	σ_m	σ_p	*Substituent*	σ_m	σ_p
CH_2CN	0·16	0·18	CHO	0·36	0·43
CH_2OR	0·02	0·03	$SO_2{\cdot}Me$	0·64	0·73
$CH_2SO_2{\cdot}CF_3$	0·29	0·31	$SO_2{\cdot}NH_2$	0·53	0·60
CH_2Cl	0·11	0·12	$SO_2{\cdot}F$	0·80	0·91
CCl_3	0·40	0·46	SF_5	0·61	0·68

What are the common structural characteristics of the fifteen substituents as a whole or divided into subsets? Examine the disposition, relative to the line, of the data for other substituents in Table 1. What do you infer from your graph? Find and interpret other relevant data for the substituents on which the line is based.

2.4. The data in the Table are for the $-\mathrm{p}K_a$ values of the protonated forms of substituted acetophenones in aqueous sulphuric acid.

$$XC_6H_4{\cdot}C(OH)^+{\cdot}Me \rightleftharpoons XC_6H_4{\cdot}CO{\cdot}Me + H^+$$

Substituent	$-\mathrm{p}K_a$	*Substituent*	$-\mathrm{p}K_a$
H	6·15	*p*-F	6·06
m-Me	6·02	*m*-Cl	7·01
p-Me	5·47	*p*-Cl	6·52
m-OMe	6·70	*m*-NO_2	7·62
p-OMe	4·81	*p*-NO_2	7·94
p-OH	4·73		

Find the type of Hammett plot which gives the best straight line. (Try σ and then σ^+.) What do you infer from it?

2.5. For a reaction mechanism consisting of a pre-equilibrium (governed by K), followed by a rate-limiting step (governed by k_2) the observed rate constant, k, is given by $k = Kk_2$. Prove that the observed ρ-value is the sum of the ρ-values for the two stages of the reaction (see p. 12).

2.6. In Table 2 the ρ-values for the ionization of substituted anilinium ions and for the benzoylation of substituted anilines are both numerically ca. 3. It might be inferred from this that in the activated complex for benzoylation (p. 11) formation of the C—N bond and the separation of the proton are almost complete. Why are such a comparison and inference invalid? Suggest a more suitable comparison of ρ-values which would shed light on the transition state in question and see if you can find any useful data in the literature (see Bibliography).

2.7. The appended Table gives rate constants ($\mathrm{l\,mol^{-1}\,sec^{-1}}$) for the alkaline hydrolysis of methyl 4-X-2,6-dimethylbenzoates in 60% aqueous dioxan at 125°C.

X	10^4k
NH_2	8·84
Me	13·5
H	17·3
Br	68·2
NO_2	352

Construct Hammett plots (taking the compound with X = H as parent), by using in turn σ, σ^0, and σ^n (Table 4). Interpret the plots and suggest an approximate value for ρ. Compare the value with data for alkaline hydrolysis of benzoates in Table 6, and try to account for the difference.

2.8. In discussing the concave-downwards Hammett plot for the acidic hydrolysis of substituted salicylidenanilines (Fig. 5) we stated that this behaviour is due to the rate-determining step not being constant throughout the reaction series. Read the original paper and identify the processes involved. [Hoffmann, J., Klicnar, J., Štěrba, V., and Večeřa, M. (1970). *Colln. Czech. chem. Commun.* **35**, 1387. The paper is in English.]

2.9. If you have access to a computer you could try multiple regression for suitable reaction series, by applying, for example, the Yukawa–Tsuno equation (8) or the dual-parameter treatment (14). The data in problem 2.4 could be examined in this way, and it is easy to find data for other reaction series in the literature.

3. The separation of polar, steric, and resonance effects

Introduction

THE occurrence of steric as well as polar effects in aliphatic systems and *ortho*-substituted aromatic systems complicates the application of correlation analysis. Fig. 3, p. 4 is an example; the plot of $\log k$ for the basic hydrolysis of ethyl alkanoates versus the $\log K_a$ values for the corresponding acids shows a wide scatter, cf. the corresponding plot for benzoate hydrolysis in Fig. 2, p. 2. Little progress in devising LFER in this area was made until R. W. Taft in 1952 devised a procedure for separating polar, steric, and resonance effects, based on an analysis of the rate coefficients of basic and acidic hydrolysis of esters. Considerable use has been made of this 'Taft analysis' and of substituent parameters developed from it.

Introduction to R. W. Taft's work

The kinetics of the formation and hydrolysis of carboxylic esters have been investigated extensively, and the operation of various structural effects influencing reactivity has long been recognized, e.g. 'steric hindrance' and 'substituent polarity'. The dominant influence of one particular factor is sometimes clear, e.g. 'steric hindrance' is responsible for the difficulty of esterifying 2,6-disubstituted benzoic acids. In many cases, however, the simultaneous operation of several factors is conceivable and a full understanding requires their separation. A method of separating polar and steric effects in ester hydrolysis was suggested by Ingold but these ideas were largely ignored until taken up by Taft (see Taft, 1956).

Taft suggested the following equation for evaluating the polar effect of a substituent R in the ester RCO_2R^1.

$$\sigma^* = [\log(k/k^0)_B - \log(k/k^0)_A]/2{\cdot}48 \tag{30}$$

σ^* is a polar substituent constant for R. The rate constants k refer to reactions of RCO_2R^1, and k^0 to those of $MeCO_2R^1$ as standard. B and A refer to basic and acidic hydrolysis, carried out with the same R^1, solvent, and temperature. The factor 2·48 puts the σ^* values on about the same scale as Hammett's σ. The equation may also be applied to *ortho*-substituted benzoic esters, o-$XC_6H_4{\cdot}CO_2R^1$, with o-$MeC_6H_4{\cdot}CO_2R^1$ as standard. The terms have the following significance: $\log(k/k^0)_B$ measures the sum of the polar, steric, and resonance effects of R (or X); $\log(k/k^0)_A$ measures the sum of the steric and resonance effects of R (or X); the difference gives the polar effect of R (or X). A further equation

$$E_s = \log(k/k^0)_A \tag{31}$$

defines a steric substituent constant, E_s, although for systems which are conjugated with CO_2R^1, E_s contains a resonance contribution.

This procedure is based on three assumptions.

(1) The relative free energy of activation, $\Delta\Delta G^{\ddagger}$, may be treated as the sum of independent contributions from polar, steric, and resonance effects.

(2) In corresponding acidic and basic reactions, the steric and resonance effects are the same.

(3) The polar effects of substituents are markedly greater in the basic than in the acidic reaction.

Assumption (1) is by no means self-evident: if the various effects interact, linear analysis is inappropriate. No progress is possible, however, unless such a simplifying assumption is made as a first approximation. Its validity is attested by the usefulness of the results obtained.

Assumption (2) lies at the heart of the analysis. The transition states (having structures closely resembling those of the carbonyl addition intermediates) for the acidic and alkaline reactions, **21** and **22**, differ by two protons.

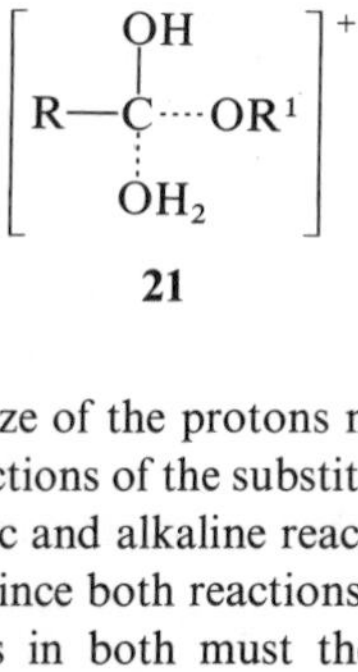

$$\left[\begin{array}{c} \mathrm{OH} \\ | \\ \mathrm{R{-}C\cdots OR^1} \\ \vdots \\ \mathrm{OH_2} \end{array}\right]^{+} \qquad \left[\begin{array}{c} \mathrm{O} \\ | \\ \mathrm{R{-}C\cdots OR^1} \\ \vdots \\ \mathrm{OH} \end{array}\right]^{-}$$

21 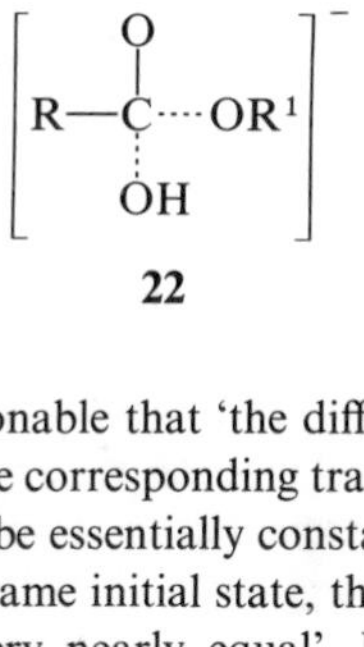 **22**

The small size of the protons makes it reasonable that 'the difference in the steric interactions of the substituent, R, in the corresponding transition states for the acidic and alkaline reactions should be essentially constant, independent of R. Since both reactions involve the same initial state, the substituent steric effects in both must therefore be very nearly equal'. Further, any resonance effect between R and CO_2R^1 should be the same in the two reactions since the initial states are the same and both transition states are effectively saturated at the carboxylate carbon. Thus the steric and resonance effects of R relative to Me should be eliminated in equation (30). We shall examine this assumption again later.

Assumption (3) is supported by the Hammett ρ-constants for the reactions of *meta*- or *para*-substituted benzoates. For basic hydrolysis, ρ-values are commonly in the range $+2{\cdot}2$ to $+2{\cdot}8$. For acidic hydrolysis or the related esterification of benzoic acids, recorded ρ-values lie in the range $0 \pm 0{\cdot}5$, and are often close to zero. Thus σ^* may be approximately scaled to σ through a factor 2·48, the average of the ρ-values recorded by Jaffé (1953) for the saponification of benzoates, and the polar contribution to E_s may be neglected.

Taft regarded the final justification of the assumptions to be the nature of the results to which they lead. He analysed a vast amount of data to give σ^*-

and E_s-values, and used these to interpret reactivity in other reactions. For some substituents the conditions regarding constancy of solvent and R^1 could not be completely met. Further, Taft used data from the esterification of carboxylic acids, RCO_2H, catalysed by hydrion in alcohols, R^1OH, to supplement those from acidic hydrolysis of RCO_2R^1 in aqueous organic media. This is justifiable in terms of the similarity of the transition states for the two processes; compare **21** and **23** which are inter-related through rapid proton transfers. For certain substituents R, for which there was plenty of

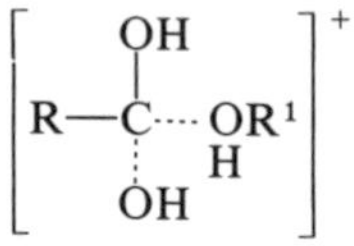

$$\left[\begin{array}{c} \text{OH} \\ | \\ \text{R—C}\cdots\text{OR}^1 \\ \vdots \quad\quad \text{H} \\ \text{OH} \end{array}\right]^+$$

23

data, it appeared that $\log(k/k^0)_A$ or $\log(k/k^0)_B$ did not depend much on solvent or R^1, and that $\log(k/k^0)_A$ was approximately the same for hydrolysis and esterification. Supposedly equivalent $\log(k/k^0)$ values at 25°C, where available, were averaged, Taft holding that 'the use of average values appears desirable on the basis that small specific effects and experimental errors will be reduced'.

Selected values of σ^* and E_s are shown in Tables 7 and 8. For a full discussion of the substituent constants for many common groups, you should consult Taft's articles. While the E_s-values in Taft's tables are always based on the original defining reaction, many of the σ^*-values are not based directly on the hydrolysis of esters but on the application of the original σ^*-values to other reactions. Table 7 contains various examples of this.

σ^*-Values

Electron-withdrawing substituents have positive values of σ^* and electron-releasing substituents have negative values. Taft regards the values of σ^* as agreeing with 'the qualitative English school theory' of the polar effect.

Branching at the α-carbon atom leads to negative values of σ^*, numerically in the order Me < Et < Pr^i < Bu^t, corresponding to the order of increase in the electron-releasing effect of these groups. The effect on σ^* of branching at the β-carbon atom or of lengthening the carbon chain beyond three atoms is relatively small; the electron-releasing effect tends to a limit. The halogens have a strong electron-withdrawing effect, with positive values of σ^* for $HalCH_2$ in the order F > Cl > Br > I, corresponding to electronegativity. The unipole $Me_3\overset{+}{N}CH_2$ is not unnaturally at the top of the list of monosubstituted methyl groups with $NC{\cdot}CH_2$ next; $NO_2{\cdot}CH_2$ would probably have $\sigma^* \sim 1{\cdot}4$, but direct experimental determination is impossible. $PhOCH_2$ is more

TABLE 7

Selected values† of σ and E_s for aliphatic systems*

R	σ*	E_s	R	σ*	E_s
Cl_3C	+2·65	−2·06	Ph_2C	+0·405‡	−1·76
Cl_2CH	+1·94	−1·54	$Cl[CH_2]_2$	+0·385§	−0·90
$Me_3\overset{+}{N}CH_2$	+1·90	—	MeCH:CH	+0·36	(−1·63)¶
MeCO	+1·65	—	$PhCH_2$	+0·215	−0·38
PhC⫶C	+1·35††	—	PhMeCH	+0·11	−1·19
$NC·CH_2$	+1·30	—	$Ph[CH_2]_2$	+0·08	−0·38
FCH_2	+1·10‡	−0·24	Me	0·00	0·00
$ClCH_2$	+1·05	−0·24	Et	−0·10	−0·07
$BrCH_2$	+1·00‡	−0·27	Pr^n	−0·115	−0·36
ICH_2	+0·85‡	−0·37	Bu^i	−0·125	−0·93
$PhOCH_2$	+0·85	−0·33	Bu^n	−0·13	−0·39
$MeCOCH_2$	+0·60	—	neopentyl	−0·165‖	−1·74
Ph	+0·60	(−2·55)¶	Pr^i	−0·19	−0·47
$HOCH_2$	+0·555††	—	cyclo-C_5H_9	−0·20‡	−0·51
$MeOCH_2$	+0·52	−0·19	Bu^s	−0·21	−1·13
H	+0·49	+1·24	Et_2CH	−0·22‖	−1·98
PhCH:CH	+0·41††	(−1·89)¶	Bu^t	−0·30	−1·54

† Taft (1956).
‡ From ionization of RCO_2H.
§ From RCO_2^- catalysis of nitramide decomposition.
¶ For significance, see text.
‖ From sulphation of alcohols.
†† From catalysis of dehydration of acetaldehyde hydrate by RCO_2H.

TABLE 8

Selected values† of σ_o^* and E_s, and related quantities for ortho-*substituents in benzoates*

X	σ_o^*	σ_o	σ_p‡	E_s	*van der Waals radius*, Å
H	—	0·00	0·00	—	—
OMe	−0·22	−0·39	−0·27	+0·99	—
Me	0·00	−0·17	−0·17	0·00	2·00
F	+0·41	+0·24	+0·06	+0·49	1·35
Cl	+0·37	+0·20	+0·23	+0·18	1·80
Br	+0·38	+0·21	+0·23	0·00	1·95
I	+0·38	+0·21	+0·28	−0·20	2·15
NO_2	+0·97	+0·80	+0·78	−0·75	—

† Taft (1956).
‡ Table 1.

electron-withdrawing than $HOCH_2$ or $MeOCH_2$. $PhCH_2$ also has a small but definite electron-withdrawing effect. Ph attached directly to CO_2R^1 shows a more powerful effect of this type. The $\alpha\beta$-unsaturated substituents are electron-withdrawing; the order of σ^* values: PhC:C $\gg$ PhCH:CH $\gg$ $PhCH_2{\cdot}CH_2$, corresponds to that of the degree of s-orbital character in the hybridization of the α-carbon atom (cf. the usual explanation of the acidity of acetylene).

The σ^*-values of the more highly substituted groups show that polar effects are approximately additive; see those for α-branched alkyl groups. The additivity is less strict when the electronic effect is large, as in Cl_2CH or Cl_3C. The 'damping effect' of a methylene group on the electron-withdrawing influence of a substituent usually involves a decremental factor of about 2·8, e.g. compare $ClCH_2$ and $Cl[CH_2]_2$.

The general order of σ_o^*-values (Table 8) corresponds very closely to that of σ_p-values. When the σ_o^*-values are changed to a σ_o-scale with hydrogen as standard by assuming that $\sigma_o = \sigma_p$ for Me, it is seen that in various other cases $\sigma_o \sim \sigma_p$. This apparently means that the polar effects of substituents operate equally from the *ortho*- and the *para*-position, but as a general rule this seems unlikely.

Values of σ^* are important for the analysis of σ-values in terms of σ_I and σ_R (Chapter 2, p. 17). For a few substituents an inductive parameter, σ', was based on the reactions of 4-substituted bicyclo[2,2,2]octane-1-carboxylic acids and esters. As mentioned on p. 24 the bicyclo-octane system provides a good model for the study of the transmission of the non-mesomeric part of the polar effect through the benzene ring. It was found that the σ'-values for X were proportional to σ^* for CH_2X. This was held to confirm that σ^*-values were true measures of the polar effect. A new inductive parameter for substituents in general was therefore defined as $\sigma_I = 0{\cdot}45\sigma^*$. In Chapter 2 we saw how this parameter has proved of considerable application to understanding the reactivities of *aromatic* compounds through a separation of inductive and resonance effects.

E_s *values*

What exactly is the steric effect which E_s measures? The transition state for acid-catalysed hydrolysis or esterification is more crowded than the initial state. When the initial state with sp^2 hybridization at the carboxy-carbon is converted into the transition state with near sp^3 hybridization, there is (i) an increase in repulsions between non-bonded atoms (a potential energy effect) and (ii) an increase in mutual interference of groups or atoms with each other's motions (a kinetic energy effect). Both of these contribute to the free energy of activation. Hence the effect on $\log k$ of replacing Me by R in principle measures a combined potential energy and kinetic energy effect. This matter is further pursued by Taft (1956); see also Shorter (1972). Note, however, that for $\alpha\beta$-unsaturated substituents the E_s-value is often governed mainly by the conjugation of R with CO_2R^1 in the initial state, and is thus not a measure of the steric effect; see entries in parenthesis in Table 7.

The E_s-values in general conform well with qualitative assessment of 'bulk'. Taft points out that σ^* and E_s are quite different functions of structure. This is shown very strikingly by the series Me, Et, Pr^i, Bu^t, in which successive methyl groups show additivity in polar effect but an increasing increment in steric effect. Also there is a substantial steric effect of β-alkyl substitution, cf. the very small polar effect. In the chloro-substituted groups the second chlorine atom produces a larger change in E_s than the first or third. When the ends of a branched structure are linked in a cyclic structure there is a substantial decrease in the steric effect, but σ^* is only slightly affected: compare cyclo-C_5H_9 with Et_2CH.

Other points of interest from the E_s values of aliphatic substituents are the similarity of E_s values for $HalCH_2$ (presumably the halogen atom can be tucked out of the way), and the strong effect of branching at the β-carbon atom in a neopentyl group.

There is doubtless some resonance contribution to the E_s values for *ortho*-substituents (Table 8), but the values seem to reflect expected steric effects. The E_s values for symmetrical *o*-X substituents parallel very well the van der Waals radii. Note that Br $\sim$ Me in steric effect, with F and Cl behaving as if smaller than Br, and I larger. OMe and OEt have much smaller steric effects than Me; no doubt they can rotate away from the reaction centre. As expected NO_2 has a large steric effect and Ph an even greater one.

Applications of the Taft parameters to the reactivity of organic compounds

There are three basic equations which deal respectively with the influences of polar effects alone, steric effects alone, and the two together. When necessary the equations are appropriately modified to include consideration of resonance effects. Each of the equations will be discussed in turn. The Taft parameters have found application in many areas and examples are given to show the variety of the applications. The use of the parameters in connection with spectroscopy and biological activity will be dealt with in later chapters.

The linear free-energy–polar-energy relationship

Taft found that the rate or equilibrium constants for a wide variety of reactions of RY conformed respectively to equation

$$\log(k/k^0) = \rho^*\sigma^* \qquad (32)$$

or the corresponding equation for equilibrium constants, where σ^* is the polar substituent constant for R and ρ^* is a reaction constant analogous to the Hammett ρ-constant. Some reactions of *ortho*-substituted aromatic systems *o*-XC_6H_4Y also obey these equations. Examples are given in Fig. 6.

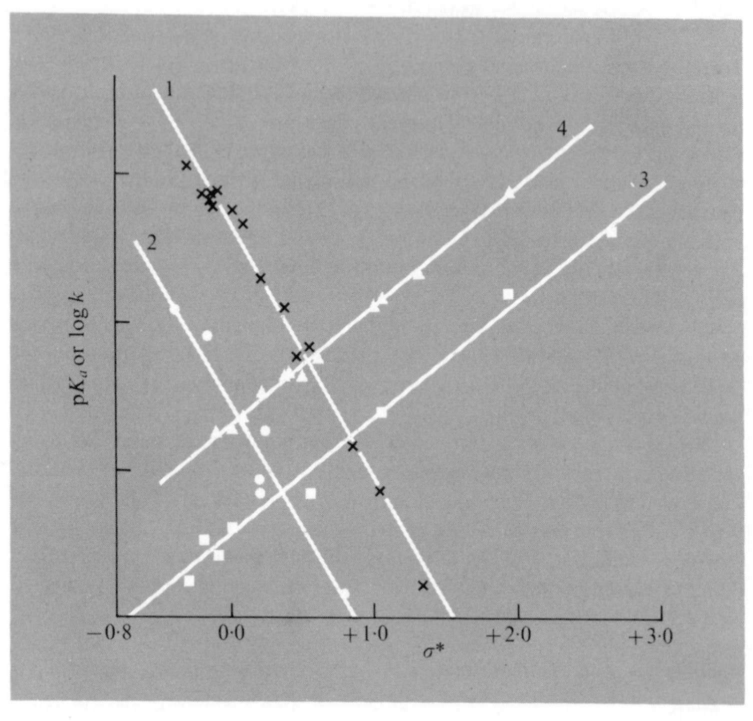

FIG. 6. Applications of the linear free-energy–polar-energy relationship. After Shorter (1970). The divisions on the ordinate are 1·00 unit of pK_a or log k apart, and the relative positions of the lines 1 to 4 with respect to the ordinate are arbitrary.

1. pK_a, aliphatic carboxylic acids, water, 25°C. 2. pK_a, *ortho*-substituted benzoic acids, water, 25°C; benzoic acid deviates markedly. 3. log k, alkaline hydrolysis of $[Co(NH_3)_5O_2CR]^{2+}$, water, 25°C.

$$[Co(NH_3)_5O_2CR]^{2+} + OH^- \longrightarrow [Co(NH_3)_5OH]^{2+} + RCO_2^-$$

4. log k, catalysis of dehydration of acetaldehyde hydrate by RCO_2H, aqueous acetone, 25°C.

$$CH_3CH(OH)_2 \longrightarrow CH_3CHO + H_2O$$

(See Shorter, 1970 for sources of data.)

Conformity to equation (32) implies that all effects other than polar remain nearly constant throughout each reaction series. Steric effects of substituents are either completely absent or approximately the same as the steric effect of a methyl group (the standard) within the range considered. A reaction series should not be held to conform to equation (32) unless substituents of a wide range of polar, steric, and resonance effects are included. The predictive power of the equation is limited: deviations may occur if substituents markedly different from those involved in the initial correlation are considered. For many of the reactions the minor role of steric effects is readily understood: some involve no change in bond angles at the reaction centre; in others the reaction centre is somewhat remote from the substituent or the reagent involved is very small.

The failure of the unsubstituted benzene system (i.e. X = H) to conform to equation (32) in most of the aromatic reaction series indicates a constant steric or resonance effect as between the unsubstituted compound and *any* of the *ortho*-substituted compounds in a given series.

Values of ρ^*, like those of ρ (Chapter 2), may help to elucidate reaction mechanisms. For example that $\rho^* = +4{\cdot}60$ for the sulphation of alcohols (33) excludes rate-determining carbonium ion formation.

$$ROH + H_2SO_4 \longrightarrow RO{\cdot}SO_2{\cdot}OH + H_2O \qquad (33)$$

The rate of oxidation of aldehyde hydrates, $RCH(OH)_2$, by aqueous chromic acid conforms to equation (32) with $\rho^* = -1{\cdot}2$. This is held to support a cyclic mechanism involving H^- transfer.

Deviations from equation (32) have been used by Taft, and subsequently by other authors, to assess other effects quantitatively.

A good example of the assessment of steric effects is provided by the application of equation (32) in a form involving enthalpies instead of free energies. Equation (35) was found applicable to the enthalpies of dissociation (kcal mol^{-1}) of the addition compounds formed between boron trimethyl and various aliphatic amines.

$$Me_3B + NR^1R^2R^3 \rightleftharpoons Me_3B{\leftarrow}NR^1R^2R^3 \qquad (34)$$

$$\Delta H = -7{\cdot}26\Sigma\sigma^* + 24{\cdot}54 \qquad (35)$$

The term $\Sigma\sigma^*$ is the sum of the σ^*-values for the groups R^1, R^2, and R^3, so that effectively the parent system is that involving NMe_3. Ammonia and straight-chain primary amines conformed well but branched-chain compounds and secondary or tertiary amines showed marked deviations. Thus ΔH(obs) for Bu^tNH_2 was $+13{\cdot}0$ kcal mol^{-1}, compared to a value of ΔH(calc) from equation (35) of $+19{\cdot}6$. The difference, $-6{\cdot}6$ kcal mol^{-1}, was attributed to steric strain in the complex. (1 kcal = 4·184 kJ.)

The study of the acidic hydrolysis of diethyl acetals of general formula $R^1R^2C(OEt)_2$ (R^1 = H or Me; R^2 variable) indicated the importance of resonance effects. Application of equation (32),

$$R^1R^2C(OEt)_2 + H_2O \xrightarrow{H^+} R^1R^2CO + 2EtOH \qquad (36)$$

with $\Sigma\sigma^*$ for R^1 and R^2, to the hydrolysis of compounds derived from non-conjugated aldehydes and ketones gave well-separated parallel straight lines for the two classes of compound, with a number of deviant points. A single straight line, however, was given by equation (37) where n is the number of α-hydrogen atoms in R^1 and R^2,

$$\log(k/k^0) = \rho^*\Sigma\sigma^* + (n-6)h \qquad (37)$$

i.e. $(n-6)$ is the decrease in the number of such atoms compared to the six in acetonal ($R^1 = R^2 = Me$), and h is a proportionality constant. The term $(n-6)h$ is interpreted as a contribution from hyperconjugation in stabilizing the transition state [resembling an oxo-carbonium ion, $R^1R^2C(OEt)^+$] relative to the initial state (which is saturated).

Substituents with $\alpha\beta$-unsaturation, e.g. phenyl, styryl, or prop-1-enyl show large positive deviations from equation (37), indicating extensive resonance interaction in the transition state, e.g. structure **24**.

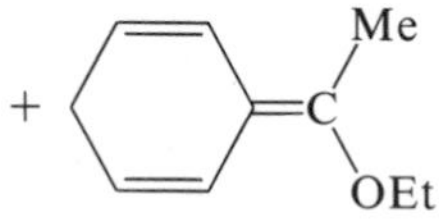

24

The transition state probably has nearly the same bond angles and distances involving the central carbon atom as the initial protonated acetal molecule, $[R^1R^2C(OEt)(HOEt)]^+$. Hence steric effects should not be large. However, very large substituents may cause steric effects, e.g. the acetal with $R^1 = Me$ and R^2 = neopentyl shows clear signs of steric acceleration, indicating a relief of strain when the tetrahedral initial state changes to the transition state in which the central carbon atom has made some progress towards trigonal hybridization.

Equation (32) has also been used to detect anchimeric assistance. The solvolysis of several tertiary alkyl chlorides in 80% ethanol at 25°C conforms to this equation, but $Me_2CCl{\cdot}CH_2I$ reacts about 740 times faster than expected, while for $Me_2CCl{\cdot}CH_2Bu^t$ the rate enhancement is about 15. The latter is reasonably attributed to steric acceleration by the neopentyl group, but the enormous effect of the much smaller CH_2I group indicates substantial anchimeric assistance† by the iodine atom in the formation of the carbonium ion **25**.

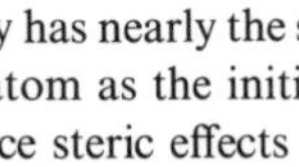

$$Me_2\overset{+}{C}\text{—}CH_2 \; (\text{bridged by I})$$

25

† i.e. a neighbouring-group effect.

Among the great variety of reaction types to which equation (32) has been applied are the following: the ionization of carboxylic acids, substituted ammonium or phosphonium ions, alcohols and thiols; the kinetic or equilibrium acidities of C—H bonds; the addition of bromine and of other reagents including free radicals to double bonds; the reactions of aldehydes and ketones; heterogeneous catalysis; polymerization; nucleophilic behaviour; and radical abstraction reactions.

The linear free-energy–steric-energy relationship

Taft found that a small number of reactions conformed to the equation:

$$\log(k/k^0) = \delta E_s \qquad (38)$$

where δ is a steric susceptibility constant. In these reactions the polar effects of substituents appear to be very small. Several of the reactions are closely related to acidic ester hydrolysis, e.g. the acidic hydrolysis of *ortho*-substituted benzamides or the alcoholysis of esters, but others are very different, e.g. the reaction of methyl iodide with 2-alkylpyridines. In addition to correlations involving $\log k$ there are some involving enthalpy of activation or of reaction. In various examples, particularly those not resembling acidic ester hydrolysis, equation (38) is of rather limited applicability; deviations occur when the range of substituents is extended.

The linear free-energy–polar- and steric-energy relationship

This is in principle the most general equation since the role of both polar and steric effects is considered. It implies that the free energy of activation is

$$\log(k/k^0) = \rho^*\sigma^* + \delta E_s \qquad (39)$$

a sum of independent contributions from polar and steric effects. An analogous equation may be written for equilibria.

Equation (39) was applied originally by Pavelich and Taft (1957) to correlate results for base-catalyzed methanolysis of (−)-menthyl esters $RCO_2C_{10}H_{19}$ in methanol.

Another application was devised by Bowden, Chapman, and Shorter (1964). The $\log K_a$ values of arylaliphatic carboxylic acids with very large groups (e.g. Ph_2CH, Ph_3C) show no clear relationship to σ^* or to E_s separately. However, multiple correlation using equation (39) is highly successful, and for 13 acids in 50% 2-n-butoxyethanol–water at 25°C yields the equation

$$\log K_a = -6{\cdot}045 + 2{\cdot}665\sigma^* + 0{\cdot}25E_s \qquad (40)$$

$$(R = 1{\cdot}000;\ s = 0{\cdot}052)$$

Since some of the E_s values were numerically as high as ca. 4, the contribution of the steric term was substantial. The acid-weakening steric effect was interpreted as steric inhibition of solvation of the carboxylate anion caused by the large arylaliphatic group.

Equation (39) has found applications to a variety of reactions from the types previously mentioned in connection with equation (32). One factor limiting its application is the difficulty in specifying steric effects in situations where more than one substituent is involved, e.g. in additions to alkenes of the type of **26**. For polar effects polysubstitution is treated in terms of $\Sigma\sigma^*$, and

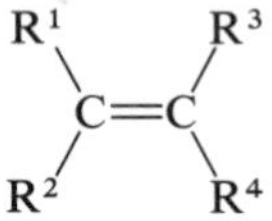

$$\begin{matrix} R^1 & & R^3 \\ & C{=}C & \\ R^2 & & R^4 \end{matrix}$$

26

this is reasonable, but the analogous ΣE_s for steric effects is not so justifiable, cf. the different behaviour of E_s and σ^* in the α-branched alkyl series, see p. 37 In spite of claims that the range of certain correlations based on equation (39) is extended by including a ΣE_s-term, the correlation treatment of such systems must be regarded as an unsolved problem (see Shorter, 1972).

As we have seen, deviations from equation (32) are often ascribed to steric effects. Many authors, however, do not succeed in incorporating the deviant systems in wider correlations involving steric parameters. It appears that in practice equation (39) is not as widely applicable as might be hoped. Certainly σ^* has found wider application than E_s.

Are there any common characteristics of the systems to which E_s is found to apply? E_s is based on a process in which a reaction centre with an sp^2 hybridized carbon atom gives rise to a transition state in which that atom is almost sp^3 hybridized. It is therefore not surprising that E_s finds its main application in systems involving the inter-relationship of sp^2 and sp^3 hybridized states, e.g. the hydrolysis of amides and esters, the alcoholysis of esters, reactions at the α-carbon atom of aliphatic nitro-compounds, and additions of molecules, ions, or radicals to C=C or C=O, including polymerization. Applications of E_s are not, however, restricted to systems with this characteristic, e.g. the use of E_s in connection with steric inhibition of solvation of carboxylate ions, as described above, or the applications to biological activity (see Chapter 7). On the other hand, E_s values seem inapplicable to the S_N2 reactions of alkyl halides with nucleophiles, which involve tetrahedrally-hybridized carbon in the initial state going to a trigonal bipyramidal transition state.

The *ortho*-effect

Steric phenomena have long been recognized as playing a major part in the peculiar effects of *ortho*-substituents. Primary steric effects of various kinds, including steric hindrance to solvation or to the approach of the reagent, and secondary steric effects have been invoked. In certain systems hydrogen-bonding and other intramolecular interactions have been postulated. The main approach to understanding the *ortho*-effect has been the attempt to separate steric effects from polar and other effects; Taft's analysis of aromatic ester hydrolysis is the best known of these attempts.

Taft's analysis for aromatic systems

Values of polar and steric parameters for *ortho*-substituents have already been discussed. Taft himself applied σ_o^* values (Me as standard) or σ_o-values (H as standard) to several reaction series by using equation (32). The parent system (X = H) tended to deviate, suggesting a difference in steric or resonance effect as between the parent compound and any of the substituted compounds. Further, ρ^* sometimes differed markedly from ρ for the corresponding *meta*- or *para*-substituted systems (e.g. $\rho^* = 1{\cdot}79$, $\rho = 1{\cdot}00$ for ionization of benzoic acids in water), although $\rho^* \sim \rho$ by definition for the alkaline hydrolysis of benzoates.

Compared with the aliphatic parameters, the Taft parameters for *ortho*-substituents have subsequently found relatively few applications, but in various studies the use of Taft's σ_o-values serves to place *ortho*-substituted compounds on the Hammett plot for *meta*- and *para*-derivatives.

Comparisons with para-*substituted systems*

The idea that *para*-substituted systems provide a basis for interpreting the *ortho*-effect goes back to 1928 when Kindler attempted to evaluate steric effects in the alkaline hydrolysis of *ortho*-substituted ethyl benzoates by comparing the behaviour of corresponding benzoates and cinnamates. The approximate equality of σ_p and Taft's σ_o-values for several substituents has encouraged various authors to use an appropriate *para*-substituent constant (σ_p, σ_p^0, σ_p^+, etc.) as a measure of the polar effect of an *ortho*-substituent. The additional structural influences constituting the *ortho*-effect may then be revealed. Similarly values of (k_o/k_p) or $\log(k_o/k_p)$ have been taken as a basis for assessing an *ortho*-effect. Shorter (1972) has assembled various examples of these approaches.

Ortho-*substituent constants*

Many authors have selected a reaction believed free from the steric effects of *ortho*-substituents and have derived a scale of σ_o-values by assuming that $\rho_o = \rho$ for this reaction. For instance, McDaniel and Brown believed that pK_a values of 2-substituted pyridinium ions provided a basis for σ_o-values. Usually, however, systems in which the reaction centre is somewhat removed from the ring have been chosen, e.g. the dissociation of arylpropiolic acids,

ArC:C·CO_2H. Values of σ_o have also been based on spectroscopic data. Charton (1969) assembled 32 sets of *ortho*-substituent constants. Selected values are given in Table 9 (see also Charton 1971).

TABLE 9

Polar constants for ortho-*substituents*†

Probable status	OMe	Me	F	Cl	Br	I	NO_2
(1) σ	−0·39	−0·17	+0·24	+0·20	+0·21	+0·21	+0·80
(2) σ	−0·13	—	+0·13	+0·24	—	—	+0·55
(3) σ	—	−0·14	+0·53	+0·83	+0·91	+0·84	—
(4) σ^0	−0·53	−0·16	+0·16	+0·31	—	—	+0·94
(5) σ^0	−0·67	−0·14	+0·23	+0·37	+0·41	+0·43	+0·97
(6) σ^+	−0·43	−0·25	—	+0·45	+0·55	—	+0·75
(7) σ^-	0·00	−0·13	+0·54	+0·68	+0·70	+0·63	+1·24
(8) σ^-	−0·37	−0·13	+0·29	+0·50	+0·55	+0·64	+1·20

† Selection by Shorter (1972).
(1) From Taft's σ_o^* values, Table 8.
(2) Ionization of phenylpropiolic acids.
(3) Ionization of benzoic acids in water, other effects having being eliminated.
(4) Gas-phase pyrolysis of isopropyl benzoates.
(5) Association of benzoic acids with 1,3-diphenylguanidine.
(6) Sundry ester pyrolyses.
(7) Ionization of phenols.
(8) P.m.r. shifts for OH of phenols.

There is often poor agreement between the various values determined for a given substituent. It is, however, important to remember that the defining processes are very varied in nature and in consequence σ_o-values may variously be the equivalent of σ, σ^0, σ^+, σ^-, etc. In some cases the defining process is probably subject to unsuspected steric or other interference from the *ortho*-substituent. If one makes allowance for all these causes of discrepancies, the agreement still seems poor, and there appear to be real difficulties in the way of determining satisfactory σ_o-values. Indeed the complexity of the influence of *ortho*-substituents on reactivity may make the search for σ_o-scales of wide applicability quite fruitless. Charton has marshalled much evidence to support such a gloomy view.

The work of M. Charton

Only the main findings can be given here. Charton (1971) has written a lengthy summary article and Shorter (1970, 1972) has given brief accounts. The work is based on the separation of the polar effect of a substituent into inductive and resonance contributions through a dual-parameter treatment involving correlation equations of the general form of (41) [cf. equation (14),

p. 19], where Q is the property being correlated ($\log k$, $\log K$, σ_o, etc.), h is the

$$Q = \alpha\sigma_I + \beta\sigma_R + h \tag{41}$$

intercept term and α and β define the relative importance of inductive and resonance effects. In each application Charton examines the significance of introducing an additional term to represent the steric effect of the substituent as a function of van der Waals radius. The main findings are as follows:
(1) Most of the proposed sets of σ_o-values may be adequately analysed in terms of σ_I- and σ_R-type parameters, but the ratio β/α varies over the range 0·3 to 1·3. Further, h varies from 0·78 to $-0{\cdot}27$ indicating the standard $\sigma_o = 0{\cdot}0$ for H is not satisfactory. Charton concludes that no completely general set of σ_o-values can be defined.
(2) A vast amount of data relating to the *ortho*-effect is analysed successfully by the above equation without the inclusion of the steric effect term. Charton concludes that steric effects play a minor role in the *ortho*-effect and that different contributions of resonance and inductive effects, i.e. β/α values account mainly for the variety of phenomena. This even applies to the E_s-values of *ortho*-substituents, which in Charton's view are thus not steric parameters at all! (Note that this comment does not apply to the E_s-values for aliphatic systems.)

Charton's work is a valuable contribution but it has been criticized on various grounds (see Shorter, 1972). At present it cannot be held to have absolutely disproved the traditional view of the importance of steric influences in the *ortho*-effect. The gloomy conclusion about σ_o-values is more justified.

The field effects of ortho-*substituents*

In Chapter 2, p. 23, we discussed the field effect model of the inductive effect. If the orientation of substituent dipoles with respect to the reaction centre is important, comparison of the influence of a given substituent in the *ortho*- and in the *para*-positions should be very revealing, since the orientations are quite different. The implications of this are incorporated in various treatments of the *ortho*-effect (Shorter, 1972). Particular consideration is given to the change in the relative transmission of the field effect through the molecular cavity and through the medium produced by varying the solvent.

The importance of dipole orientation is shown strikingly in certain systems in which a normal substituent effect is greatly diminished or even reversed. For example, 8-substituted-1-naphthoic acids **27** have a very unusual disposition of substituent and functional group; the 8-chloro and 8-methyl acid have almost identical pK_a values.

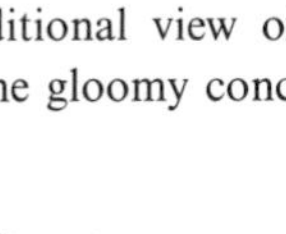

X CO_2H

27

More recent developments and critique of the Taft analysis

Much of our account of the Taft analysis may appear as a considerable success story in a previously intractable area. A note of warning was, however, sounded in connection with the *ortho*-effect and there are other *caveats* which must be mentioned.

The nature of the steric parameter

We have already discussed the kind of steric effect which E_s is considered to measure and have seen that for substituents conjugated with a carboxylic function, E_s contains a substantial resonance component. Taft recognized that even for substituents incapable of normal conjugation, there might be a contribution to E_s from a resonance interaction, i.e. a hyperconjugative effect of α-hydrogen atoms. This comment seems to have escaped most organic chemists, but in the view of Hancock,

$$\overset{+}{H}CH_2{=}C(OR^1)(O^-)$$

28

Meyers, and Yager (1961) this effect should be allowed for in a 'corrected steric substituent constant', E_s^c, as in equation (42),

$$E_s = E_s^c + h(n-3) \qquad (42)$$

where h is a reaction constant for hyperconjugation and n is the number of α-hydrogen atoms. Quantum mechanical calculations were used as a basis for taking h as $-0{\cdot}306$ at 35°C. Selected E_s^c-values are in Table 10. Hancock showed that the use of E_s^c-values could lead to significant improvement in the correlation of the results for certain

TABLE 10

Taft's† E_s and Hancock's‡ E_s^c for R *in* RCO_2R^1

R	E_s	E_s^c
Me	0·00	0·00
Et	−0·07	−0·38
Pr^i	−0·47	−1·08
Bu^t	−1·54	−2·46
Pr^n	−0·36	−0·67
Bu^n	−0·39	−0·70
Bu^i	−0·93	−1·24

† Table 7.
‡ Hancock, Meyers, and Yager (1961).

reactions. The following relationships were found to hold for the saponification of nine esters, RCO_2Me, in 40% aqueous dioxan at 35°C. (Rate constants in this section

$$\log k = 1{\cdot}31 + 1{\cdot}54\sigma^* + 0{\cdot}709E_s \qquad (43)$$

$$\log k = 1{\cdot}36 + 1{\cdot}48\sigma^* + 0{\cdot}471E_s^c \qquad (44)$$

$$\log k = 1{\cdot}25 + 1{\cdot}75\sigma^* + 0{\cdot}848E_s^c - 0{\cdot}383(n-3) \qquad (45)$$

are in $l\,mol^{-1}\,min^{-1}$.) For equation (43) $R = 0{\cdot}992$ and $s = 0{\cdot}076$, i.e. the correlation is fairly good because hyperconjugation is involved in E_s and in $\log k$. Equation (44) is poor ($R = 0{\cdot}970$ and $s = 0{\cdot}149$) because in E_s^c hyperconjugation has been eliminated. Equation (45) is excellent ($R = 0{\cdot}998$ and $s = 0{\cdot}043$) because it incorporates proper consideration of both steric and hyperconjugative effects.

Hancock has also shown the importance of 'change in the six-number'. M. S. Newman proposed that the number of atoms in position 6, from the carbonyl oxygen as 1, has a large influence on the steric effect. When a given group is considered both as R and R^1 in RCO_2R^1 there may be a change in the six-number, $\Delta 6$, as between R and R^1. $\Delta 6$ is the six-number of a given substituent as R minus its six-number as R^1. Thus for Me, Et, Pr^i, and Bu^t, $\Delta 6 = 0, -3, -6$, and -9 respectively. Hancock uses $\Delta 6$ as another structural parameter. For the saponification of 9 acetates, $MeCO_2R^1$ in 40% dioxan at 35°C, equation (46) holds, with $R = 0{\cdot}980$ and $s = 0{\cdot}161$, the

$$\log k = 1{\cdot}40 + 1{\cdot}34\sigma^* + 0{\cdot}730E_s^c \qquad (46)$$

E_s^c-values being for the substituents as R. When $\Delta 6$ is included, equation (47) holds, with $R = 0{\cdot}997$ and $s = 0{\cdot}070$, which is a much improved correlation.

$$\log k = 1{\cdot}35 + 0{\cdot}688\sigma^* + 0{\cdot}664E_s^c + 0{\cdot}0477\Delta 6 \qquad (47)$$

A few authors have used Hancock's parameters but in general his ideas have not been taken up by organic chemists.

Modified steric parameters have also been developed by Palm, who considers the contribution of both C—H and C—C hyperconjugation to E_s. Shorter (1972) has summarized some of Palm's work.

The significance of σ for alkyl groups*

There seems little doubt that σ^*-values measure the polar effects of substituents when these are substantial. σ^*-Values show good relationships to various electrical properties, e.g. electronegativity values, ionization potentials, dipole moments, and certain parameters of molecular orbital theory (Shorter, 1972). The significance of small σ^*-values may be questioned, and this applies to the values for alkyl groups which lie mainly between 0 and $-0{\cdot}3$. Such small values might arise from an imperfect cancelling of steric effects in the Taft analysis (p. 33). We shall consider this matter particularly later (p. 48). There have, moreover, been other contributions casting doubt on the status of σ^*-values for alkyl groups. We can only summarize these briefly; for a more extended summary see Shorter (1972).

T. L. Brown (1959) discussed the possible role of polarizability effects in the influence of alkyl groups on reactivity. He concluded that such effects were so variable that the general applicability of σ^*-values based on ester reactions should be questioned. Ritchie and Sager (1964) advanced subtle arguments to suggest that σ^*-values for alkyl groups are not consistent with those of other substituents and are in fact bogus. Shorter (1972) has suggested that there may be an underlying fallacy in the argument, but both these pieces of work have acquired new interest in view of recent discoveries as to the way in which alkyl groups influence gas-phase reactivity.

Modern techniques based on mass spectrometry enable gas-phase acidities and basicities to be measured. For the *acidities* of a series of amines the order

$$Bu^tNH_2 \geqslant Pr^iNH_2 > EtNH_2 > MeNH_2 > NH_3 \qquad (48a)$$

was observed (Brauman and Blair 1969). Interpreted conventionally in terms of the inductive effect this suggests that the *electron-attracting* power of alkyl groups *increases* with chain length and branching. On the other hand for gas-phase *basicities* orders such as

$$Me_3N > Me_2NH > MeNH_2 > NH_3 \qquad (48b)$$

are observed (Munson 1965), corresponding to a 'normal' electron-releasing inductive effect of methyl groups.

It appears that in the gas-phase alkyl groups stabilize both cations and anions. In solution the stabilization of anions is not observed. Hence the behaviour of alkyl groups as manifested in reactions in solution must be very much influenced by interactions with the solvent. Thus it may well be naive to suppose the σ^*-values for alkyl groups are a simple measure of an inherent polar effect.

Critique of the Taft analysis

In this final section, in a sense, we return to the beginning and re-examine the fundamental assumptions of Taft's procedure. We can only do this very briefly; again see Shorter (1972).

Before we comment on the assumptions it is appropriate to mention one practical matter. Taft's analysis is commonly said to be a treatment of acidic and basic ester hydrolysis. In practice it is largely a treatment of *acid-catalysed esterification* and basic ester hydrolysis, for there is still a paucity of data relating to acidic and basic ester hydrolysis in the same solvent.

Criticism of Taft's treatment has been mainly directed at assumption (2) concerning the equality of steric effects as between corresponding acidic and basic reactions. Attention has been focused on the neglect of the role of the solvent: since the transition states of the acidic and basic reactions carry opposite charges, it is unlikely that the solvation patterns will be so similar that the steric interactions in the two systems will be exactly the same. Various lines of evidence, summarized by Shorter (1972), bear on this matter but it cannot be said that any definitive position has yet been reached.

Assumption (3) regarding the negligibility of a polar effect on $\log(k/k^0)_A$ has also been questioned. In particular polar effects are by no means small in acid-catalysed esterification in methanol, a reaction on which many recorded E_s-values depend very heavily.

There seems little tendency to question assumption (1) regarding the approximate separation of polar, steric, and resonance effects in a linear way. A widespread disbelief in it would stultify study of this field. Possible contributions from interaction between various effects are sometimes considered. The inclusion of cross-terms in correlations frequently improves them, but often their physical significance is obscure.

PROBLEMS

3.1. (*a*) Find as many examples as possible from Table 7 to show the damping effect of a CH_2-group on σ^*. Suggest a σ^*-value for CF_3 on the basis of the following values: CF_3CH_2, $+0{\cdot}92$; $CF_3[CH_2]_2$, $+0{\cdot}32$; $CF_3[CH_2]_3$, $+0{\cdot}12$.

(*b*) The σ^* value for Ph(OH)CH is $+0{\cdot}765$. How does this compare with the value you can predict from data in Table 7?

(*c*) The σ^* value for Me_3SiCH_2 is $-0{\cdot}26$ and for 9-fluorenyl **A** is $+0{\cdot}50$.

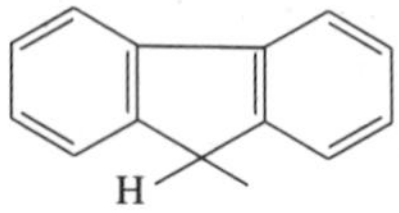

H

A

Comment on these values by comparing each of them with relevant data in Table 7.

3.2. It is much more difficult to understand the patterns of E_s values than of σ^*-values. Try to develop your own understanding of the E_s data in Table 7, supplemented by data from Taft (1956) and Bowden, K., Chapman, N. B., and Shorter, J. (1963). *J. chem. Soc.* 5239. These references contain much relevant discussion. [The latter source contains σ^* and E_s data for arylaliphatic groups, e.g. **A**.]

3.3. The Table gives rate constants ($l\,mol^{-1}\,sec^{-1}$) for the acidic hydrolysis of amides $XCH_2{\cdot}CO{\cdot}NH_2$ in water at 75°C.

XCH_2	10^4k	XCH_2	10^4k
Et	12·0	cyclo-$C_6H_{11}CH_2$	1·24
Pr^n	5·99	$MeOCH_2$	8·98
Bu^n	5·93	$ClCH_2$	12·1
Bu^i	1·29	$BrCH_2$	11·1
$PhCH_2$	5·19	neopentyl	0·193

Plot the log k-values against (*a*) σ^* and (*b*) E_s for XCH_2. (The necessary values of σ^* and E_s are mainly in Table 7. For cyclo-$C_6H_{11}CH_2$ use $\sigma^* = -0{\cdot}06$ and $E_s = -0{\cdot}98$.) Which plot gives the better straight line? Consider the significance of this plot and any deviations.

3.4. (*a*) The Table gives rate constants ($l\,mol^{-1}\,min^{-1}$) for the addition of bromine to olefins $RCH{:}CH_2$ under fixed conditions.

R	k	R	k
Et	2900	CH_2OEt	89·8
Me	1840	$CH_2{\cdot}O{\cdot}COMe$	9·03
$Ph[CH_2]_2$	985	CH_2Cl	2·20
$[CH_2]_3O{\cdot}CO{\cdot}CF_3$	353	CH_2CN	0·875
$PhCH_2$	312		

Plot log k against σ^*. (Some of the σ^* values are in Table 7; other values are $[CH_2]_3O{\cdot}CO{\cdot}CF_3$, 0·26; CH_2OEt, 0·495; $CH_2{\cdot}O{\cdot}COMe$, 0·87.) Draw the best straight line and calculate ρ^*. Comment on the success of the correlation.

(*b*) Examine the disposition of the following data in relation to the line you have drawn.

R	k	R	k
Pr^n	2090	Bu^s	925
Pr^i	1700	Bu^t	802
Bu^n	1990	neopentyl	346
Bu^i	994		

Can you find any correlation between the observed deviations and E_s values (Table 7)?

(*c*) Attempt to examine the applicability of the above correlation to the following rate constants: $CH_2{:}CH_2$, 30·3; $MeCH{:}CH_2$, 1840; $Me_2C{:}CH_2$, $1{\cdot}64 \times 10^5$; $Me_2C{:}CHMe$, 4×10^6; $Me_2C{:}CMe_2$, $5{\cdot}5 \times 10^7$.

4. Applications to spectroscopy

Introduction

IN Chapters 2 and 3 we saw how polar substituent constants can be applied to the understanding of organic reactivity. Can these constants be applied to any of the physical properties of organic compounds? The various types of spectroscopy provide the most fruitful area for investigating this. For many years authors have attempted to correlate frequencies or intensities of absorption bands in optical spectra with substituent constants in infrared spectroscopy with considerable success. Less progress has been made for ultraviolet absorption spectra. Substituent constants have been applied to chemical shifts or coupling constants in nuclear magnetic resonance spectra, mainly those associated with ^{1}H or ^{19}F. In this area there have been some notable successes but many obscurities remain. Finally there is the application of substituent constants in mass spectrometry, as yet relatively undeveloped.

In this chapter we shall explore these various areas in turn, with emphasis on infrared and nuclear magnetic resonance spectroscopy. For more detailed spectroscopic background, see e.g. Fleming and Williams (1966), McLauchlan (1972).

Infrared spectra

A non-linear molecule of n atoms has $3n-6$ modes of internal vibration. Each vibrational mode which involves variations in molecular dipole moment is excited by infrared radiation of the same frequency, which is thus absorbed from an infrared beam. Strictly speaking each vibrational mode involves the whole molecule, but in a large molecule some of the modes are essentially associated with the vibrations of particular bonds. These may be called localized vibrations and they involve either the stretching or the deformation of bonds. It is only for a localized vibration that we can expect to relate the frequency or intensity of the infrared absorption to the electrical properties of substituents in the molecule. This restriction limits us mainly to vibrations of high force constant such as C—H, N—H, O—H, C=C, C≡C, and C=O stretching modes.

The positions of infrared absorption bands are commonly expressed in wave-number units (cm^{-1}) but are often loosely called 'frequencies'. The true unit of frequency is, however, the herz (reciprocal second). We shall not be much concerned with the quantitative specification of band position and shall use the term frequency in a general sense, and denote it by ν. The same Greek letter will also be used to designate a vibrational mode: thus ν_{NH_2}(sym) means the symmetrical stretching mode of an amino group.

There are various possible ways for expressing the intensity of an infrared absorption. Intensities have been measured satisfactorily only recently. At one time the usual practice was to record the *extinction coefficient*, ε, determined from the peak height in a spectrum of a solution by equation (49). In equation (49), c is the concentration of the solution, l the path length, and T

$$\varepsilon = (1/cl)\log(100/T) \tag{49}$$

is the percentage transmission. Peak heights tend to be a function of the instrument and mode of working. It is now the practice to evaluate the *integrated intensity*, A, by integrating equation (50) over the whole band. [See Katritzky and Topsom (1972) for further discussion of methods of expressing intensity.]

$$A = (1/cl)\int \log(100/T)\mathrm{d}\nu \tag{50}$$

Besides the limitation to localized vibrations, there is one other which should be noted. Infrared bands are often very sensitive to solvent effects (see Chapter 5). In the search for relationships between the characteristics of bands and substituent constants, such solvent effects should be minimized. Gas-phase measurements would be ideal but the use of solutions in an inert solvent, e.g. a hydrocarbon, is more practicable. Naturally the same solvent must be used throughout each series of related compounds.

Infrared frequencies

The absorption frequency is directly proportional to the energy quantum of the relevant vibrational mode, $h\nu$. In developing correlation equations for infrared frequencies it is therefore logical to replace the $\log k$ or $\log K$ terms of linear free-energy relationships by ν. Thus the Hammett equation may be rewritten as in (51), where ν_0 is the statistical quantity corresponding to the

$$\nu = \nu_0 + \rho\sigma \tag{51}$$

frequency for the parent member of the series.

Jaffé's review (1953) summarized applications of equation (51) to data for several different series of aromatic compounds. The data were mainly for the ν_{CO} absorption of substituted benzoyl chlorides, benzoic acids, acetophenones, etc. and quite good correlations were reported. Many later papers have presented similar correlations between ν and σ for a variety of absorptions in various series of aromatic compounds. The application of the σ^+ and σ^- scales to such data has also been examined. Unfortunately the choice of substituents has not always been very suitable for testing the relevance of the different scales of polar parameters, and many obscurities remain.

There is, however, a fair amount of evidence that the correlation of a particular series of results may require σ^+, σ, or σ^- depending on the electronic demand of the vibration concerned. Thus the vibrations of strongly $+R$ substituents such as NO_2 or CN, e.g. ν_{NO_2}(asym) or ν_{CN}, often require

σ^+, while those of strongly $-R$ groups such as OH or NH_2, e.g. ν_{OH} or ν_{NH_2}(asym), often require σ^-. For the vibrations of groups whose resonance interaction with the ring is feeble, such as CO_2H, e.g. ν_{OH}, σ is satisfactory.

As for chemical reactivity (see p. 15), adherence to the 'duality of σ-values may be criticized, and an attempt made to get away from this by using the dual-parameter equation, which now takes the form (52). The possible

$$\nu = \nu_0 + \rho_I\sigma_I + \rho_R\sigma_R \tag{52}$$

relevance of σ_R^0, σ_R^+, σ_R^-, and σ_R (from benzoic acid) must be considered (see p. 19). Depending on the vibration being examined, different resonance parameters give the best correlation and the ratio $\rho_R/\rho_I = \lambda$ measuring the blending of resonance and inductive effects varies quite widely.

It must be emphasized, however, that it is not difficult to find many series of infrared frequencies from aromatic compounds for which sensible correlations have not so far been devised and there is no apparent reason for the difficulty in many cases.

Various correlations of the absorption frequencies of aliphatic compounds with σ^* have been made. Thus ν_{OH} values for eleven substituted acetic acids, and ν_{CO} values of RCOMe (with R as a variety of hetero-atom groups) are well correlated with σ^*.

Infrared intensities

The intensity of an infrared absorption, A, is proportional to the square of the *dipole transition moment*, i.e. the rate of change of dipole moment with respect to the relevant coordinate. The logical quantity to use in correlations with σ-values is therefore the square root of the intensity, $A^{\frac{1}{2}}$. Before this was realised, however, successful correlations involving $\log A$ (based on a mistaken analogy with the $\log k$ or $\log K$ term of the Hammett equation) or A had been reported. The success was probably spurious and due to the small range of substituents used. Also, until fairly recently, measured intensities were not very reliable. For infrared intensities the Hammett equation is therefore now used in the form (53). As with frequencies there is the possibility that certain

$$A^{\frac{1}{2}} = A_0^{\frac{1}{2}} + \rho\sigma \tag{53}$$

series may require σ^+ or σ^- rather than σ, but the distinction is often obscured by inadequate choice of substituents.

Approximate correlations of $A^{\frac{1}{2}}$ with σ-values were obtained a long time ago for various aromatic series, e.g. appropriate side-chain vibrations in benzaldehydes, phenols, anilines, benzonitriles, acetanilides, and benzoates. The ρ-values depend on the nature of the vibrating species, e.g. the ρ-value for the ν_{OH}-vibration of phenols is positive, but the values for the ν_{CO}-vibration of benzoates and the ν_{CN}-vibration of benzonitriles are negative; the actual ρ-values show a rough relationship to the σ-value of the vibrating group.

Exner has shown that $A^{\frac{1}{2}}$ values for an extensive series of benzonitriles (including twenty-two *para*-substituted compounds) follow σ^+. $A^{\frac{1}{2}}$ values in certain aliphatic series have been successfully correlated with σ^*, e.g. the $\nu_{C=C}$ vibration of ethylenes of the type $CH_2{:}CH[CH_2]_nX$ with $n = 1, 2, 3$, and the ν_{OH} vibration of R_3SiOH.

Various more sophisticated types of correlation for infrared intensities have been established in recent years. Schmid (1966) has measured the intensity of the ν_{CH}-vibrations in many mono- and di-substituted benzenes and in substituted pyridines. The intensities are related to the σ_I values of the substituents and the most successful equation is of the form:

$$A = a\sigma_I^2 - b\sigma_I + e \tag{54}$$

where *a*, *b*, and *e* are constants for a particular substituent pattern. The work appears to confirm the reality of the separation of σ-values into inductive and resonance components.

Katritzky and Topsom have measured integrated intensities for the ν_{16} and ν_{13} ring vibrations of substituted benzenes. This work can only be summarized here: see Katritzky and Topsom (1972). The intensities appear to be governed by the resonance effects of substituents, and for the ν_{16} ring-stretching vibration of monosubstituted benzenes, the following equation applies:

$$A = 17600(\sigma_R^0)^2 + 100 \tag{55}$$

The intercept term (100) is due to an overtone band. Fig. 7 represents graphically some of the results. The line provides the means of determining previously unknown σ_R^0-values (although the sign is not given) and of presenting a set of self-consistent σ_R^0-values. Analogous results are obtained for the ν_{13} band, and the treatment has also been extended to $\nu_{C=C}$ of substituted ethylenes and $\nu_{C\equiv C}$ of substituted acetylenes. When the $A^{\frac{1}{2}}$-values for monosubstituted benzenes are treated by the dual-parameter method (σ_I and σ_R^0) the blending constant λ is about 24, indicating the dominance of the resonance effect.

Ultraviolet spectra

Correlation analysis of data from u.v. spectra presents a number of difficulties. Electronic excitation is commonly accompanied by changes in vibrational energy, so that a u.v. absorption band tends to be broad and to possess a vibrational fine-structure. The peak maximum (at ν_{max}) usually corresponds to simultaneous changes in electronic and vibrational energy, and the vibrational contribution may well be changed by introducing substituents. For ideal correlation analysis the $0 \rightarrow 0$ transition (molecule in vibrational ground state in both the electronic ground and excited state) should be identified but in practice correlation of u.v. frequencies has usually to be done with values of ν_{max}. Further, a u.v. spectrum often has several absorptions so that peak positions may be difficult to determine and it may not be easy to identify the corresponding peaks in the spectra of a series of related compounds. Also the introduction of a substituent may cause a radical change in the nature of the molecular

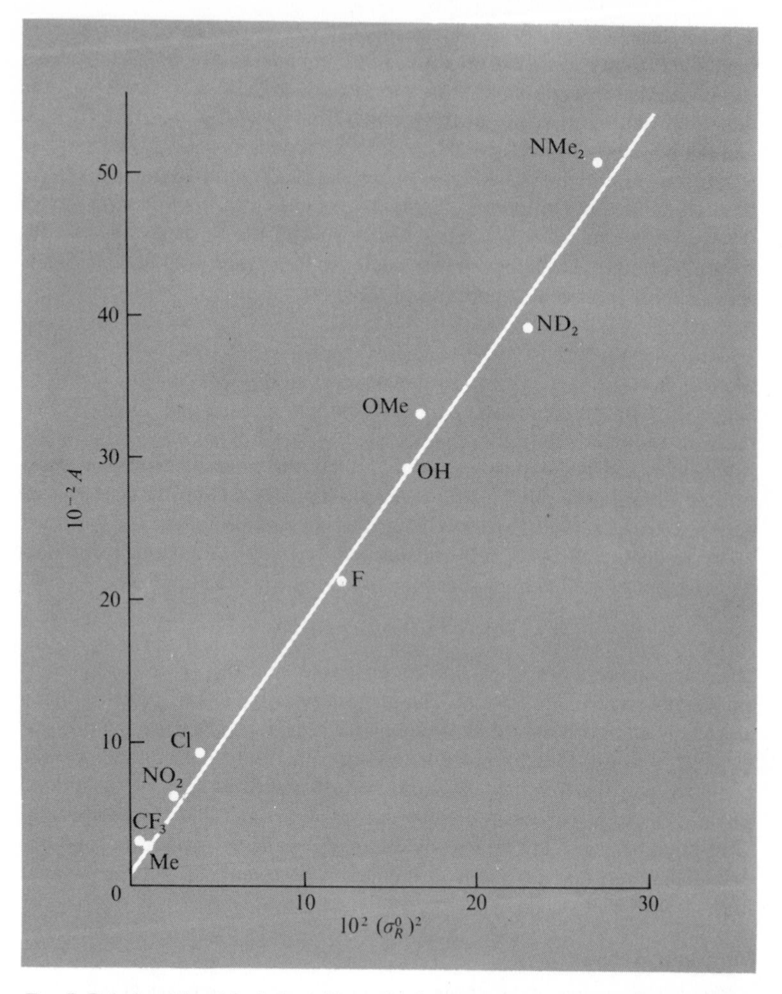

FIG. 7. Relationship of the infrared intensity for the ν_{16} ring vibrations of substituted benzenes to $(\sigma_R^0)^2$ values of substituents. After Brownlee, Hutchinson, Katritzky, Tidwell, and Topsom (1968). The compounds were in solution in carbon tetrachloride.

orbitals or may contribute its own u.v. absorption and the attempt to correlate the spectra of a series of compounds is then quite pointless. The determination of meaningful u.v. absorption intensities is made difficult by the same factors and in practice the best that can be done is to try to evaluate integrated intensities by the same equation as is used for infrared spectra.

After this gloomy introduction it will not be surprising that the achievements of correlation analysis in u.v. spectroscopy are somewhat limited. However, this may be due partly to unfortunate choices of series of compounds and substituents, and it is likely that more carefully designed and executed work will bear fruit.

The spectra of monosubstituted benzenes have three main bands (wavelengths for benzene itself, in hexane, are given in parentheses): the secondary band (254 nm), the primary band (203·5 nm), the second primary band (184 nm). The first two of these are 'forbidden' by 'symmetry' for the $0 \rightarrow 0$ transition of benzene itself but the bands are shown by benzene because of the molecular vibrations. Most work on the influence of substituents has involved the primary band and the secondary band. All substituents, irrespective of their electronic character, shift the maximum to lower frequency. If the substituents are classified into $-R$ or $+R$ groups, the following orders are obtained for the size of the shift in the case of the primary band:

$$-R\text{: Me} < \text{Cl} < \text{OH} < \text{OMe} < \text{NH}_2 < \text{O}^-$$

$$+R\text{: CN} < \text{CO}_2\text{H} < \text{COMe} < \text{CHO} < \text{NO}_2$$

The resonance effect thus appears to be very important. Limited correlations of ν_{max} with $(\sigma_p - \sigma_m)$ or σ_R have been claimed, but with a wide range of substituents the dual-parameter treatment finds no significant correlation.

It could be argued that by considering the u.v. absorption of benzene we are starting with an over-complicated system. The study of a suitable chromophore in an aliphatic system might be more fruitful. Most work has been done on the carbonyl group, which has four ultraviolet absorptions. The one most studied is the $n \rightarrow \pi^*$ (singlet) absorption, which is at 280 nm for acetone. For aliphatic ketones expressions such as

$$\nu = \nu_0 + \rho^*\Sigma\sigma^* + n_h h_h + n_c h_c \tag{56}$$

have been suggested, where n_h and n_c are the number of hydrogen and of carbon atoms attached to α-carbon atoms respectively, and h_h and h_c are constants. Presumably these terms describe hyperconjugative effects. Studies involving a wider range of substituents have led to expression (57). σ_R^+ is used instead of σ_R for substituents

$$\nu - \nu_0 = 10^4\Sigma(\sigma_I - \sigma_R) \tag{57}$$

joined directly to the CO group. The equation reveals an opposing influence of inductive and resonance effects; possibly inductive electron-withdrawal lowers the energy of the ground state, while resonance electron-donation increases the energy of the excited state.

Intensity data have usually been treated in the form of 'spectroscopic moments' or 'oscillator strengths', but effectively the relationship of the square root of the intensity to σ-values has been examined. There has been some success with the dual-parameter equation for the secondary band of monosubstituted benzenes, with a blending factor of about 4, indicating the importance of the resonance effect.

Nuclear magnetic resonance spectra

Certain atomic nuclei, by virtue of nuclear spin, behave as magnets. The most important nuclei of this type are ^1H, ^{13}C, and ^{19}F, which have nuclear spin of $\frac{1}{2}$ in units of $h/2\pi$. In this brief introduction it will be convenient to consider ^1H and thus present the basis of *proton magnetic resonance*.

When a proton is placed in a uniform magnetic field, the nuclear magnet may be aligned either with the field (low-energy orientation) or against the field (high-energy orientation). Transition between these two levels may be induced by electromagnetic radiation of the correct frequency. For an applied magnetic field of about 10^4 gauss, the radiation required is in the radiofrequency range of the spectrum, about 40 MHz. The resonance frequency, ν, is related to the local field, H, experienced by the proton by the equation

$$\nu = \gamma H/2\pi \tag{58}$$

where γ is a constant, the *gyromagnetic ratio*. When the proton is part of a molecule the local field it experiences will be less than the applied field, H_0, because of the magnetic shielding effect of the electrons around the nucleus. (The applied field causes the electrons to circulate and produce a secondary field opposed to the applied field in the vicinity of the nucleus.) The shielding may be represented by a shielding parameter, σ (an unfortunate choice of symbol from the standpoint of this book), so that

$$H = H_0(1-\sigma) \tag{59}$$

and

$$\nu = \gamma H_0(1-\sigma)/2\pi \tag{60}$$

Protons with different shielding parameters, i.e. in different electronic environments, may be successively brought into resonance by varying either ν or H_0; and both modes of operation are used.

The position of a proton resonance is conventionally measured relative to that of the twelve equivalent protons in the molecule of tetramethylsilane (TMS). The absorption frequency is normally measured directly in Hz and may be converted into field independent units by dividing by the operating frequency of the instrument (in MHz) to obtain the *chemical shift*, δ, in parts per million, or ppm.

The local field experienced by a proton depends not only on its electronic environment but also on the spin-orientation of neighbouring protons. This effect, which is mediated by the spins of bonding electrons, is known as *spin–spin coupling*. Thus in the unit

$$\overset{(a)}{CH_2X}-\overset{(b)}{CHX_2}$$

29

the proton (*b*) will experience three slightly different local fields depending on whether the protons (*a*) are oriented (i) both with the applied field, (ii) both against the applied field, or (iii) one with and one against the field. The resonance for proton (*b*) occurs at three slightly different frequencies (a 'triplet') separated by the coupling constant, J. For specifying the chemical shift, the centre of the band is used.

^1H chemical shifts

In so far as proton chemical shifts depend on the immediate electronic environment of the proton, electron-attracting or -repelling substituents might be expected to influence them. There is much evidence that this is so. Thus in monosubstituted benzenes, electron-attracting substituents displace the chemical shifts of the ring protons to higher frequencies (deshielding), while electron-repelling substituents produce the opposite effect (shielding). Some correlations with chemical reactivity parameters might well be expected, but it must be remembered that n.m.r. is concerned with the ground state of the molecule, cf. chemical reactions. Many authors have tried to relate proton chemical shifts in aromatic compounds to Hammett σ-values. A selection of results is given in Table 11.

TABLE 11

Hammett correlations in p.m.r.†

Compound	*Solvent*	$-\rho$	*n*	*r*
m- or *p*-$XC_6H_4\mathbf{H}$	CCl_4	0·652	27	0·918
m-$XC_6H_4\mathbf{H}$	CCl_4	0·540	14	0·797
m- or *p*-$XC_6H_4C\mathbf{H}_3$	CCl_4	0·198	20	0·866
m-$XC_6H_4C\mathbf{H}_3$	CCl_4	0·209	10	0·807
m- or *p*-$XC_6H_4C⋮C\mathbf{H}$	CCl_4	0·334	11	0·977
m- or *p*-$XC_6H_4C\mathbf{H}_2O_2C{\cdot}CH_3$	CCl_4	0·172	9	0·898
m- or *p*-$XC_6H_4CH_2O_2C{\cdot}C\mathbf{H}_3$	CCl_4	0·093	9	0·973
m- or *p*-$XC_6H_4O\mathbf{H}$	DMSO	1·27‡	28	0·962
m- or *p*-$XC_6H_4N\mathbf{H}_2$	DMSO	1·36‡	12	0·980

† Mainly from Tribble and Traynham (1972). Ordinary σ_m or σ_p values used except as indicated.
‡ σ^- values used.

The ρ-values are all negative since substituents of positive σ-value lead to deshielding. It is important to state the solvent involved since the susceptibility of chemical shifts to substituents is solvent-dependent, cf. ρ-values for rates and equilibria. Some of the correlations seem quite successful as judged by the correlation coefficient. This applies particularly to correlations for protons somewhat remote from the benzene ring. For ring protons and those separated from the ring by one other atom the correlations are often not so good, although this is not necessarily so. Indeed, for measurements using dimethyl sulphoxide (DMSO) as solvent, the correlations for hydroxy protons in phenols and amino protons in anilines are very striking. (The DMSO forms very strong hydrogen bonds to the hydroxy or amino group. This prevents disturbances due to self-association, which occur in solutions in hydrocarbons, CCl_4, etc., leading to a large deshielding effect and reducing the susceptibility to substituent effects.) In spite of these successes it seems certain that substituents in a benzene ring do not influence proton chemical shifts and chemical reactivity in strictly analogous ways. This is shown very strikingly by certain systems for which there is poor correlation for the influence of *meta*-substituents on proton chemical shifts. A substituent may contribute its own magnetic effect to modify the field experienced by the proton some distance away. This has no parallel in reactivity parameters. The magnetic effect is due to the intrinsic magnetic anisotropy of most bonds, and its magnitude is highly dependent on distance and angle.

Benzene rings have a special magnetic effect. When a benzene ring is placed in a magnetic field the π-electrons are caused to circulate. This *ring current* produces a secondary field which reinforces the applied field in the region of the ring protons and gives a deshielding effect. Substituents on the ring, which are capable of resonance interaction with the ring, interfere with the π-electrons and hence the ring current.

Attempts have been made to correct observed chemical shifts for these effects, and improved Hammett correlations have resulted. Presumably the satisfactory straightforward Hammett correlations already noted arise when the proton under study is insulated by distance or in some other way from the effects of substituent anisotropy and ring current.

There are relationships between the chemical shifts of *ortho*-substituted compounds (dissolved in highly basic solvents) and Taft's σ_o constants, e.g. the amino protons in anilines, and the carboxy protons in benzoic acids. For phenols in DMSO the chemical shifts for *ortho*-substituted compounds parallel those for the *para*-substituted compounds, and the system has been suggested for the measurement of σ_o^- values (see p. 44 and Tribble and Traynham, 1972).

In aliphatic systems attempts have been made to correlate proton chemical shifts with σ^*, but the situation is rarely simple. Thus chemical shifts for CH_3 or CH_2 protons of acetates or succinates respectively have been correlated with σ^*-values for the alkyl group of the alkoxy-substituents of the ester. The correlations were not very good, deviations being particularly marked for bulky substituents. The deviations were explained in terms of steric inhibition of resonance involving the canonical form $X{-}C\overset{-}{O}{=}\overset{+}{O}{-}R$. Proton chemical shifts in a variety of aliphatic systems have been corrected empirically for substituent anisotropy effects and excellent correlations with the sum of the σ^*-values for the substituents involved have then been obtained.

We cannot here explore in detail the influence of substituents on coupling constants. This is a complicated topic, since a variety of effects can occur. However, some Hammett correlations have been found, e.g. for geminal coupling constants ($^2J_{HH}$) for the CH_2 protons in *meta*- or *para*-substituted benzyloxytetrahydropyrans (**30**) (see Tribble and Traynham, 1972).

O—CH_2— X

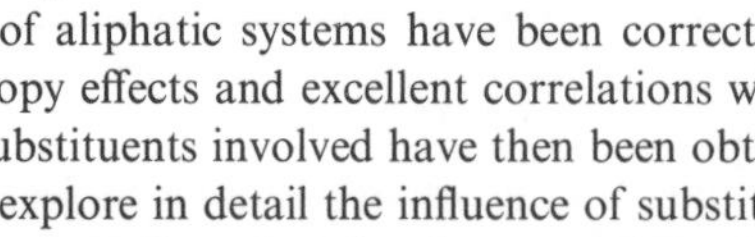

30

^{19}F chemical shifts

According to the theory of n.m.r., the chemical shifts for ^{19}F and other relatively heavy nuclei should be less sensitive to many of the perturbations which operate for protons. The shielding effect of the electrons in the vicinity of the nucleus should be dominant. Hence it might be hoped that ^{19}F chemical shifts would provide a good probe for electronic disturbances caused by substituents and that the results might be related to reactivity parameters. To some extent this has been realized. The relative ^{19}F chemical shifts in *meta*- or *para*-substituted fluorobenzenes are not, however, related in a straightforward way to Hammett σ-values. Several years ago Taft, Price, Fox, Lewis, Andersen, and Davis (1963) found a linear relationship between the relative chemical shifts in *meta*-substituted fluorobenzenes and the σ_I values of the substituents. Further, the differences between the chemical shifts in corresponding *meta*- and *para*-substituted compounds were linearly related to σ_R^0. Apparently the *meta*-shifts are predominantly influenced by the inductive effect and the *para*-shifts by the resonance effect. There was no obvious explanation for this but Taft took it as evidence for the reality of the separation of Hammett substituent constants into inductive and resonance contributions. More recently Brownlee and Taft (1970) have applied the dual-parameter equation to ^{19}F chemical shifts of fluorobenzenes, and have obtained the following relationships:

$$\mathcal{S}_{\mathrm{H}}^{m-\mathrm{X}} = 5{\cdot}3\sigma_I + 0{\cdot}6\sigma_R^0 \tag{61}$$

$$\mathcal{S}_{\mathrm{H}}^{p-\mathrm{X}} = 6{\cdot}1\sigma_I + 30{\cdot}4\sigma_R^0 \tag{62}$$

Recently Taft's interpretation of ^{19}F chemical shifts in aromatic systems has been criticized by Dewar, who has compared the ^{19}F chemical shifts of 3′- and 4′-substituted 4-fluorobiphenyls, 3″-substituted 4-fluoroterphenyls, and substituted 1- and 2-fluoronaphthalenes with those of substituted fluorobenzenes. Dewar and his colleagues (see e.g. Adcock and Dewar, 1967) have concluded that 'the effects of substituents on ^{19}F chemical shifts are qualitatively different from their effects on physical and/or chemical properties of other side-chains; attempts to explain both sets of phenomena in terms of a common inductive/resonance scheme are therefore incorrect'. The n.m.r. behaviour of *meta*-substituted fluorobenzenes is particularly 'out of step'. The matter has not yet been resolved, which is unfortunate since many σ_I and σ_R values have been based on Taft's correlations for ^{19}F chemical shifts in fluorobenzenes, and it is possible that such values are suspect.

^{13}C chemical shifts

Until recently the study of these has been somewhat difficult. As for ^{19}F chemical shifts the relevant structural factors should be simpler than for p.m.r. Some correlations with Hammett constants have been reported, and downfield shifts resulting from increased branching of alkyl groups in ketones and alcohols have been interpreted in terms of an increased electron-*attracting* influence of alkyl groups (cf. Chapter 3), but this interpretation has been criticized.

Mass spectra

In a mass spectrometer an organic vapour is bombarded by electrons of energy *ca.* 70 eV. The main initial process is the ionization of the molecule to give a molecular ion. This requires energy in the region of 10 eV. Surplus

$$AB \longrightarrow AB^{+\cdot} + e^- \quad (63)$$

electronic energy may be used in breaking bonds in the molecular ion to produce fragment ions and neutral species.

$$AB^{+\cdot} \longrightarrow A^+ + B^\cdot \quad (64)$$

or

$$AB^{+\cdot} \longrightarrow B^+ + A^\cdot \quad (65)$$

If the initial molecule is complex a considerable variety of fragment ions may be produced. The ions are separated according to their ratios of charge to mass. The photographic record of the mass spectrum gives the relative intensities and charge:mass ratios of the ions.

The principal fragmentations undergone by the molecular ion are those which give rise to the most stable carbonium ions, radicals, or (sometimes) neutral molecules. Only the charged species are actually detected in the mass spectrometer. In so far as the introduction of substituents is likely to affect the stabilities of the molecular ion and of the fragmentation products, some relationship of intensities to suitable substituent constants might be hoped for, and has been found. Most examples relate to aromatic systems and involve the Hammett equation.

If the intensities of the ions are represented by chemical formulae in square brackets, then the useful quantity for the fragment A^+ is its intensity relative to that of the molecular ion $AB^{+\cdot}$, and is denoted by Z.

$$Z = \frac{[A^+]}{[AB^{+\cdot}]} \quad (66)$$

In a rather over-simplified way Z may be regarded as a measure of the rate constant for the breakdown of $AB^{+\cdot}$ to A^+ and $B^\cdot$. In the usual way for Hammett correlations we consider data for a parent system (Z_0) and a substituted system (Z). The appropriate quantity to correlate with σ-values is $\log(Z/Z_0)$, i.e. we use the Hammett equation in the form (67).

$$\log(Z/Z_0) = \rho\sigma \quad (67)$$

One of the best known examples is provided by the mass spectra of *meta*- and *para*-substituted benzophenones. The relevant fragmentation process is

$$C_6H_5{\cdot}CO{\cdot}C_6H_4X^{+\cdot} \longrightarrow C_6H_5{\cdot}CO^+ + C_6H_4X^{\cdot} \quad (68)$$

so that

$$Z = \frac{[C_6H_5{\cdot}CO^+]}{[C_6H_5{\cdot}CO{\cdot}C_6H_4X^{+\cdot}]} \quad (69)$$

The Hammett plot is shown in Fig. 8, with $\rho = 1{\cdot}01$. Somewhat similar

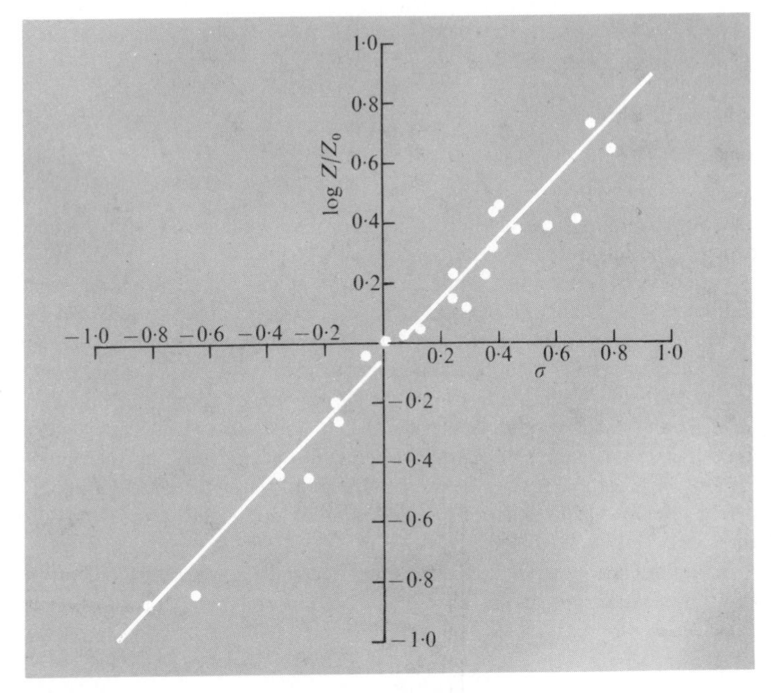

FIG. 8. Hammett plot for fragmentation of the molecular ion of substituted benzophenones to give the benzoyl cation. After Bursey (1972).

examples are summarized in Table 12. Since the correlations involve the application of parameters from reactions in solution to reactions in the gas phase, the conformity to the Hammett equation is quite pleasing, even though there are a number of deviant points. The ρ-values are positive because localization of positive charge on the unsubstituted ring is facilitated by electron-attracting groups.

The alternative cleavage reactions in which the positive charge is left on the substituted ring, e.g.

$$C_6H_5{\cdot}CO{\cdot}C_6H_4X^{+\cdot} \longrightarrow C_6H_5{\cdot}CO^{\cdot} + C_6H_4X^+ \quad (70)$$

TABLE 12

Hammett correlations of ion intensities in mass spectra†

Process	ρ
$XC_6H_4{\cdot}CO{\cdot}C_6H_5^{+\cdot} \longrightarrow C_6H_5CO^+$	1·01
$XC_6H_4{\cdot}CO{\cdot}C_6H_5^{+\cdot} \longrightarrow C_6H_5^+$	0·0
$XC_6H_4{\cdot}CO{\cdot}CH_3^{+\cdot} \longrightarrow CH_3CO^+$	0·78
$XC_6H_4{\cdot}CO_2{\cdot}CH_3^{+\cdot} \longrightarrow CH_3OCO^+$	0·67
$XC_6H_4{\cdot}N{:}N{\cdot}C_6H_5^{+\cdot} \longrightarrow C_6H_5N_2^+$	1·05

† After Bursey (1968).

give poor Hammett correlations. This is probably connected with the tendency of the substituted ion to decompose further.

The above considerations, while satisfactory up to a point, are greatly oversimplified. The intensities of product ions in mass spectra are governed by various factors in a complicated way. The decomposition of the molecular ion constitutes a set of competitive and consecutive unimolecular gas reactions. The intensity of a given product ion depends not only on the rate at which it is formed but also on the rate at which it is decomposed. Also when the molecular ion can undergo a number of parallel fragmentation reactions, the intensity of the ion produced in one of these will be influenced by the rates of all the competing processes. The situation requires a much more detailed analysis, see Bursey (1972).

Energy quantities associated with mass spectra, i.e. ionization potentials and appearance potentials, have also been correlated with substituent constants.

PROBLEMS

4.1. The Table gives values for the antisymmetrical stretching frequency of SO_2 in substituted benzene sulphonyl chlorides $XC_6H_4{\cdot}SO_2Cl$. [Data from Weigman, H. -J. and Malewski, G. (1966). *Spectrochim. Acta* **22**, 1045.]

Substituent	ν (cm^{-1})	*Substituent*	ν (cm^{-1})
H	1387·5	*m*-NO_2	1394·5
m-Me	1385·1	*p*-NO_2	1392·7
p-Me	1385·5	*m*-OMe	1386·0
m-Cl	1390·6	*p*-OMe	1384·6
p-Cl	1389·9	*m*-CO_2H	1391·5
m-Br	1390·5	*p*-CO_2H	1390·8
p-Br	1390·1	*p*-Ph	1386·4

In most cases the uncertainty in ν is given as $\pm 0{\cdot}5\ cm^{-1}$. Plot ν against σ (Table 1) and comment on the graph. Investigate also the possible relevance of σ^+ and σ^-

(Table 3). The following values for *ortho*-substituted compounds are also given.

Substituent	ν (cm^{-1})
o-Me	1379·6
o-Cl	1388·9
o-Br	1387·9
o-NO_2	1393·6
o-OMe	1382·2

Examine the applicability of Taft's σ_o-values (Table 9).

4.2. The Table gives the chemical shifts (ppm, relative to benzene) for the *p*-^{1}H and 4-^{13}C nuclei in substituted benzenes (^{1}H measurements in cyclohexane as solvent, ^{13}C measurements on the pure liquid). [Data are from Spiesecke, H. and Schneider, W. G. (1961). *J. chem. Phys.* **35**, 731.]

Substituent	^{1}H-*shift*	^{13}C-*shift*
F	0·217	4·4
Cl	0·117	2·0
Br	0·030	1·0
I	0·033	0·4
Me	—	2·8
OMe	0·367	8·1
NH_2	0·625	9·5
NMe_2	0·615	11·8
CHO	−0·275	−6·0
COMe	—	−4·2
NO_2	−0·333	−6·0
H	0·000	0·0

Plot (*a*) the ^{1}H-shift against the ^{13}C-shift (as far as possible), (*b*) the ^{13}C shift against (i) σ_p, and (ii) σ_R^0. Comment on the plots. (The σ-values are in Tables 1 and 5, except for the following: *p*-CHO, $\sigma_p = 0·43$; $\sigma_R^0 = 0·18$; *p*-NMe_2, $\sigma_R^0 = -0·54$.)

4.3. The ionization potential of a molecule is the energy required to produce the molecular ion from the ground state of the molecule. A mass spectrometer can be used to measure ionization potentials but they are better measured by *photo-electron spectroscopy*. The Table gives ionization potentials (eV) for alkyl bromides. [Data from Hashmall, J. A. and Heilbronner, E. (1970). *Angew. Chemie* (International Edn. in English) **9**, 305.] A lone-pair electron is removed in the process:

$$RBr \longrightarrow RBr^{+\bullet} + e^-$$

R	*I*	R	*I*
H	11·84	Pr^i	10·26
Me	10·69	Bu^t	10·09
Et	10·45	Bu^i	10·25
Pr^n	10·33	neopentyl	10·19
Bu^n	10·28		

The mean error is stated to be $\pm 0·015$ eV. Plot *I* against σ^* (Table 7) and comment on the graph.

5. Solvent effects

Introduction: the concept of solvent polarity

SOLVENT effects on organic reactivity have been studied for more than a century, but even the most superficial understanding of them could not precede some considerable knowledge of reaction mechanisms, molecular structure, the transition state theory, and intermolecular forces. The importance of electrostatic forces has long been recognized and organic chemists have usually attempted to understand solvent effects in terms of the *polarity* of the solvent. The work of Hughes and Ingold was a notable landmark in this effort (p. 65).

The concept of solvent polarity is easily grasped qualitatively: it relates to promoting the separation of unlike charges or the approach of like charges, but it is difficult to define precisely and even more difficult to express quantitatively. Reichardt (1965) states that the polarity of a solvent 'is determined by its solvation behaviour, which in turn depends on the action of intermolecular forces (Coulomb, directional, inductive, dispersion, and charge-transfer forces, as well as hydrogen-bonding forces) between the solvent and the solute'. Attempts to express it quantitatively commonly involve properties such as dielectric constant, dipole moment, or refractive index. We shall see later that this procedure is often very inadequate (p. 68).

Physical chemists have tried to develop quantitative treatments of solvent effects on rates in terms of models of the initial state and activated complex, with electrostatic interactions involving ions and/or dipoles as appropriate, such interactions being moderated by the dielectric behaviour of the solvent (p. 66). Since dielectric constant is often chosen by organic chemists as a measure of solvent polarity, the approaches of the two classes of chemist are not really very different and both tend to over-emphasize the importance of dielectric constant.

Another over-emphasized feature has been work with mixtures of water and an organic solvent. Up to a point this has been inevitable since many well-known reactions (e.g. ester hydrolysis) are conveniently studied in such media. The properties both of water itself and of aqueous solutions of non-electrolytes are still far from understood. Among possible complications are preferential solvation of a solute by one of the components of a mixed solvent, and the intervention of chemical equilibria involving the solute and the components of the solvent. The interpretation of solvent effects in aqueous organic mixtures is therefore often superficial, even when it is apparently successful. The same comment applies to mixtures of two organic solvents.

The inadequacy of defining solvent polarity in terms of individual physical characteristics such as dielectric constant, has stimulated various attempts to base a scale of polarity on some convenient solvent-sensitive reference process (p. 70). However, it is probable that the search for a unique general scale of solvent polarity is pointless. Solvent–solute interactions are so complex that a much more detailed analysis is required, with a variety of parameters to express different aspects of solvent–solute interaction (p. 74).

Qualitative theory of the influence of the solvent on reaction rate

Before we discuss the quantitative treatment of solvent effects, and in particular the application of correlation analysis, a brief account of the qualitative approach associated with Hughes and Ingold (1935) is appropriate. For a more detailed account see Ingold (1969).

The basic ideas may be stated as follows:

(*a*) The rates of reactions in which ionic charges arise or are compressed into a smaller space during passage of the reactants from the initial state to the transition state increase with increase of the polarity of the medium surrounding the reactants and the activated complex.

(*b*) Conversely if ionic charges disappear or are dispersed throughout a larger space during the passage to the transition state, the reaction is retarded by an increase in the polarity of the medium.

(*c*) Reactions in which ionic charges arise or disappear are much more strongly influenced by the polarity of the medium than those in which the activation process simply involves charge diffusion.

The charge-type to which the reaction belongs is thus of fundamental importance. Hughes and Ingold developed a systematic classification of reactions in terms of charge-type. We now examine two well-known simple examples.

The solvolysis of t-butyl chloride in hydroxylic solvents proceeds by the S_N1 mechanism, in which the rate-determining step is the heterolysis of the C—Cl bond.

$$\mathrm{Me_3CCl} \rightleftharpoons \left[\mathrm{Me_3\overset{\delta+}{C}\cdots\overset{\delta-}{Cl}}\right]^{\ddagger} \rightleftharpoons \mathrm{Me_3\overset{+}{C}} + \mathrm{Cl^-} \xrightarrow[\text{with solvent}]{\text{Reaction}} \text{products} \qquad (71)$$

This is a reaction in which ionic charges arise during the formation of the activated complex. It therefore proceeds much more rapidly in water, which is a very polar solvent (high dielectric constant, strong solvation of both anions and cations) than in the alkanols, which are much less polar. The relative rate constant as between water and ethanol is about 10^5.

Reactions between nucleophilic ions Y^- and primary alkyl halides RX proceed by the S_N2 mechanism, in which there is synchronous breaking of the C—X bond and formation of the C—Y bond.

$$\mathrm{Y^-} + \mathrm{RX} \rightleftharpoons \left[\mathrm{\overset{\delta_1-}{Y}\cdots R\cdots\overset{\delta_2-}{X}}\right]^{\ddagger} \rightleftharpoons \mathrm{YR} + \mathrm{X^-} \qquad (72)$$

This is a type of reaction in which charge is dispersed throughout a larger space during the activation process. It is therefore retarded by an increase in the polarity of the solvent, but the change on going from ethanol to water is not so dramatic as in the previous example: depending on the nucleophile and the substrate, retardation may be by a factor of up to about 50.

The charge-type of the reaction is of fundamental importance because this governs the extent of interaction between the solvent and the reactants in the initial state and in the transition state. When the passage to the transition state involves the creation of ionic charges or a concentration of charge, the activated complex will be solvated more strongly than the original reactants. An increase in the polarity of the solvent will therefore favour the transition state relative to the initial state. Conversely when the activation process involves the disappearance of ionic charges or a dispersal of charge, the reactants will be solvated more strongly in the initial state than in the transition state. An increase in the polarity of the solvent will then favour the initial state relative to the transition state.

The Hughes and Ingold treatment of solvent effects is certainly an over-simplified approach, particularly in its dealing with aqueous organic solvent mixtures, and in assuming the dominance of enthalpy rather than entropy effects. However it works well over a wide area, and its emphasis on the importance of charge distribution in reactants and in the activated complex is undoubtedly correct.

The treatment of solvent effects on reaction rate in terms of dielectric constant

For this purpose bimolecular reactions may be classified as dipole–dipole, ion–dipole, or ion–ion reactions. The last-mentioned category is rare in organic chemistry; we will deal in turn with the first two types. (See Laidler, 1963 for details.)

Dipole–dipole reactions

Kirkwood derived expression (73) for the change in the free energy of a molecule of dipole moment μ and radius r when transferred from a medium of unit dielectric constant to one of dielectric constant ε.

$$\Delta G = -(\mu^2/r^3)(\varepsilon - 1)/(2\varepsilon + 1) \tag{73}$$

The application of (73) and the appropriate equation of transition state theory [equation (2), p. 3] to a reaction between dipolar molecules A and B gives equation (74) where ‡ designates the parameters of the activated complex and

$$\ln k = \ln k' - \frac{N_A}{RT} \cdot \frac{\varepsilon - 1}{2\varepsilon + 1}\left[\frac{\mu_A^2}{r_A^3} + \frac{\mu_B^2}{r_B^3} - \frac{\mu_\ddagger^2}{r_\ddagger^3}\right] \tag{74}$$

k' is the rate constant for a medium of unit dielectric constant. (It is assumed that only electrostatic forces need be considered.)

The formation of the activated complex is often accompanied by considerable charge separation, in which case $\mu_{\ddagger}$ is very much greater than μ_A or μ_B. Equation (74) then predicts that the plot of $\ln k$ against $(\varepsilon-1)/(2\varepsilon+1)$ will be a straight line of positive slope. If values for all the quantities in square brackets could be suggested, equation (74) would predict the actual slope. Most of the quantities are somewhat ill-defined, particularly those for the transition state, and the best that can be done quantitatively is to work backwards and show that the observed slope fits in with 'reasonable' values.

For many reactions in binary solvent mixtures, whose dielectric constants change with composition, good linear plots of $\log k$ versus $(\varepsilon-1)/(2\varepsilon+1)$ are obtained. This applies, for example, to the Menschutkin reaction (75) in mixtures of acetone with dioxan, benzene, or tetrahydrofuran. However, each

$$Et_3N + EtI \longrightarrow Et_4\overset{+}{N} + I^- \tag{75}$$

binary solvent system leads to its own straight line of distinctive slope, which is not at all in accord with equation (74). This clearly indicates the limitations of dielectric constant and the importance of specific interactions between solvent and the various solutes. Presumably the Kirkwood function of dielectric constant 'works' for each binary solvent system because the solutes are effectively saturated with the specific interactions over the composition range employed.

For the same reaction in a series of one-component solvents, there is a general tendency for $\log k$ to increase with $(\varepsilon-1)/(2\varepsilon+1)$ but there are many anomalies; see Fig. 9. No doubt specific solvent–solute interactions vary widely over this range of solvents. The *polarizability* of the solvent molecule may be important. With the halogenobenzenes the reaction is fastest in the member of lowest dielectric constant, iodobenzene, the most polarizable of the halogenobenzenes.

Ion–dipole reactions

For an ion the equation corresponding to (73) for a dipole is (76), where Ze

$$\Delta G = \frac{-Z^2e^2}{r}\left(\frac{\varepsilon - 1}{2\varepsilon}\right) \tag{76}$$

is the charge and r is the radius of the ion. The application of (76) and transition state theory to an ion–dipole reaction gives (77). The electrostatic free-energy contribution of the reactant dipole is neglected. Since $r_{\ddagger} > r$, a linear

$$\ln k = \ln k' + \frac{N_A Z^2 e^2}{RT}\left(\frac{\varepsilon - 1}{2\varepsilon}\right)\left(\frac{1}{r_{\ddagger}} - \frac{1}{r}\right) \tag{77}$$

relationship between $\ln k$ and $(\varepsilon-1)/2\varepsilon$ of negative slope is predicted.

The data available for testing the equation are mainly for reactions in

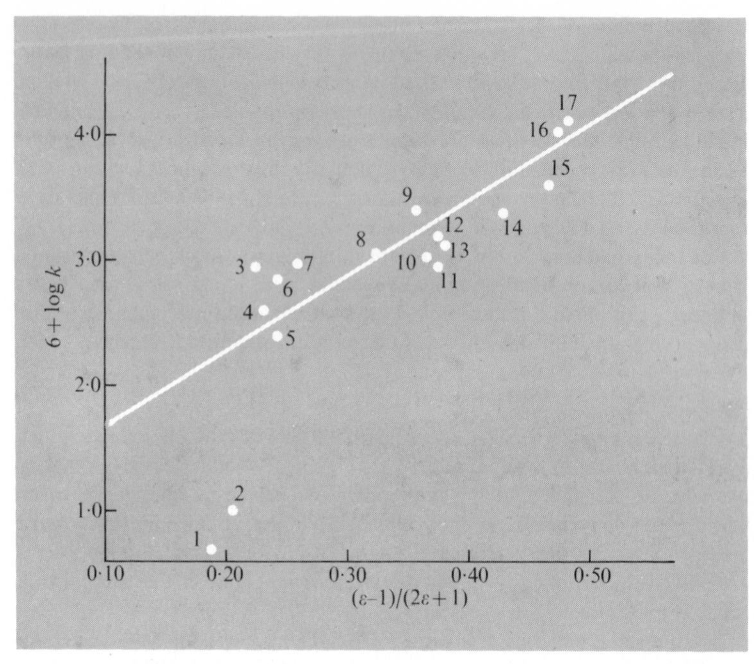

FIG. 9. Attempted correlation of log k (l mol^{-1} sec^{-1}, 100°C) for reaction between triethylamine and ethyl iodide with the Kirkwood function of dielectric constant. After Wiberg (1963). Data from Grimm, Ruf, and Wolff (1931). Key to solvents: 1. hexane; 2. cyclohexane; 3. dioxan; 4. benzene; 5. toluene; 6. *p*-dichlorobenzene; 7. diphenylmethane; 8. diphenyl ether; 9. iodobenzene; 10. *m*-dichlorobenzene; 11. fluorobenzene; 12. bromobenzene; 13. chlorobenzene; 14. *o*-dichlorobenzene; 15. acetone; 16. benzonitrile; 17. nitrobenzene.

aqueous organic mixtures. The linear relationship is sometimes observed, but there are various difficulties in dealing with this kind of reaction. For instance any reaction involving OH^- ions in alcohol–water mixtures will be complicated by the equilibrium

$$OH^- + ROH \rightleftharpoons OR^- + H_2O \qquad (78)$$

In certain solvents the reactant ion may be subject to pairing with its counter-ion, e.g. Na^+OH^-, particularly when ε is low. Part of the observed solvent effect may arise because the ion-pair is much less reactive than the free ion.

The limitations of dielectric constant

The application of bulk dielectric constant to interactions on a molecular scale is in principle inadmissible. The solvent cannot be 'between' the ions or dipoles, which are about to react, in the same sense as the solvent is between the plates of a condenser in the measurement of ε. For binary solvent mixtures the objection is even stronger: solvent–solute interactions may tend to concentrate one component in the vicinity of the molecules of one of the reactants. The local composition of the solvent will then not be the same as the overall composition. Thus the local dielectric constant cannot possibly be the same as that corresponding to the given overall composition. Nevertheless, as we have seen, the application of ε is attended with some success, particularly in the realm of mixed solvents. We shall see later how more refined theories of solvent effects retain ε as a measure of the influence of the solvent on general electrostatic interactions on the molecular scale.

Possibly ε should be replaced by n_D^2 as a measure of 'dynamic' dielectric constant; n_D^2 is the square of the refractive index for sodium light.† Dielectric constants are measured in a low frequency field in which dipoles can be orientated to follow the field; refractive index involves interaction with a high frequency field, and is a measure of *polarizability*, whereas dielectric constant includes *polarization*. The rapidity of molecular processes might make n_D^2 more relevant to solvent effects than ε. Equations (74) and (77) can be rewritten with n_D^2 replacing ε. The amended equations are usually less successful than the original, but n_D^2 may be used to supplement ε (see p. 74).

The strongest criticism of the use of ε as a measure of polarity is that it does not express the *specific* nature of interactions between solvent molecules and reactants or activated complex. Solvent molecules may have highly localized centres which interact specifically with localized centres in solutes. Such interactions are not adequately represented by radius, overall charge, and dipole moment. A protic solvent (alcohol, carboxylic acid, etc.) may be able to form hydrogen bonds with suitable negative centres in the solute. Also any solvent with unshared electron pairs may enter into donor–acceptor interaction with electron-poor centres in the solute. This is particularly important for the so-called *dipolar aprotic solvents*, e.g. acetone, dimethyl sulphoxide, and *NN*-dimethylformamide. We now give two examples which show the importance of hydrogen bonding. The S_N2 reaction between aniline and ω-bromoacetophenone:

$$PhNH_2 + PhCO\cdot CH_2Br \longrightarrow Ph\overset{+}{N}H_2\cdot CH_2\cdot COPh + Br^- \qquad (79)$$

involves the formation of a highly polar activated complex. Protic solvents interact very strongly with the $Br^{\delta-}$ end of the activated complex through hydrogen-bonding, thereby stabilizing the complex and facilitating reaction. Thus the rate constant for methanol is about six times that for nitrobenzene, even though the dielectric constants of the solvents are almost the same.

Anion–dipole reactions are much faster in dipolar aprotic solvents than in

† n_D^2, rather than n_D, is used because of the connection between the dielectric constant of a substance and the square of its refractive index which is part of J. Clerk Maxwell's electromagnetic theory of light. Values for sodium light are used because these are available for numerous liquids.

alcohols. Thus the reaction between *p*-fluoronitrobenzene and azide ion:

$$p\text{-}NO_2C_6H_4F + N_3^- \longrightarrow p\text{-}NO_2C_6H_4N_3 + F^- \qquad (80)$$

is much faster in DMF than in methanol (by a factor $>10^4$), even though the dielectric constants differ only slightly. The azide ion is heavily solvated by hydrogen bonding in methanol, and much of the solvation shell must be removed in forming the activated complex. In DMF the azide ion is in a much more exposed and therefore reactive condition.

The recognition of the limitations of dielectric constant has led various authors to attempt to dispense with it entirely in favour of *solvent parameters* based on actual chemical or physical processes. In this way the originators of each solvent parameter hope to allow for the complexity of solute–solvent interactions. We shall now examine some of the best-known solvent parameters. [See Reichardt (1965) and Koppel and Palm (1972).]

Solvent parameters

The majority of these are based either on linear free-energy relationships for chemical processes or on solvatochromic† shifts in electronic spectra. We consider first examples in the former category.

The Grunwald and Winstein Y, *and related parameters*

These authors (1948) suggested treating solvent effects on rates in terms of equation (81), similar in form to the Hammett equation. $\log k$ refers to a given reaction in a given solvent, $\log k^0$ to the same reaction in 80% v/v aqueous

$$\log k = \log k^0 + m\mathrm{Y} \qquad (81)$$

ethanol as a standard solvent, Y is a parameter characteristic of the given solvent, and m is a parameter characteristic of the given reaction, which measures its susceptibility to changes in solvent; the analogy of Y and m with σ and ρ respectively is obvious. Scales of Y and m were established by taking $\mathrm{Y} = 0$ for 80% ethanol, and selecting the solvolysis of t-butyl chloride at 25°C as a standard reaction for which m is defined as 1·000. Y values are known for various one-component solvents (mainly alcohols) and for various mixtures of organic solvents with water or a second organic solvent. Selected values of Y are in Table 13. From the nature of the standard reaction, Y values will clearly provide some measure of the ability of the solvent to solvate ions. It was therefore hoped that Y values might be a more generally applicable measure of solvent polarity than dielectric constants.

In a large number of cases the Grunwald–Winstein equation is fairly successful, and in some cases $\log k$ appears to be correlated better with Y than

† Solvatochromic shift and solvatochromism mean the displacement (to higher or lower frequency) of a spectral band, as caused by interaction of the absorbing molecule with molecules of solvent.

TABLE 13

Solvent parameters†

Solvent	ε	Y	*Z*	E_T
Dioxan	2·21	—	—	36·0
Carbon tetrachloride	2·23	—	—	32·5
Chloroform	4·7	—	63·2	39·1
Acetic acid	6·2	−1·639	79·2	51·9‡
t-Butyl alcohol	12·2	−3·26	71·3	43·9
Pyridine	12·3	—	64·0	40·2
i-Propyl alcohol	18·3	−2·73	76·3	48·6
Acetone	20·5	—	65·7	42·2
Ethanol	24·3	−2·033	79·6	51·9
Methanol	32·7	−1·090	83·6	55·5
NN-dimethylformamide	36·7	—	68·5	43·8
Methyl cyanide	37·5	—	71·3	46·0
Dimethyl sulphoxide	46·6	—	71·1	45·0
Water	78·5	3·493	94·6	63·1
Formamide	109·5	0·604	83·3	56·6
Ethanol : water, 80 : 20	—	0·000	84·8	53·6
Acetone : water, 80 : 20	—	−0·673	80·7	52·2

† Data mainly from Reichardt (1965), and measured at 25°C.
‡ Based on a correlation between E_T and *Z*.

with functions of dielectric constant. Good linear relationships between $\log k$ and Y are shown by the solvolyses of various tertiary halides and secondary sulphonates, i.e. reactions which proceed by an S_N1 mechanism, like the standard reaction. The situation for S_N2 reactions (e.g. primary halide solvolysis) or for reactions of borderline mechanism (e.g. secondary halide solvolysis) is less satisfactory. If $\log k$ is plotted against Y for the solvolysis of diphenylmethyl chloride in aqueous ethanol, aqueous acetone, or aqueous dioxan, each binary solvent system gives a different straight line of characteristic slope. If the Grunwald–Winstein equation were strictly obeyed all the points should lie on a single straight line. Such observations may indicate that the rates of reaction depend not only on the 'ionizing power' of the solvent (considered to be measured by Y) but also on the 'nucleophilicity' of the solvent. Such a property may clearly be relevant for reactions with substantial S_N2 characteristics. Winstein, Grunwald, and Jones (1951) suggested that the equation (82) might be applicable, with Y being a measure of the ionizing

$$\mathrm{d}\ln k = \left(\frac{\partial \ln k}{\partial \mathrm{Y}}\right)_{\mathrm{N}} \mathrm{dY} + \left(\frac{\partial \ln k}{\partial \mathrm{N}}\right)_{\mathrm{Y}} \mathrm{dN} \qquad (82)$$

power and N a measure of the nucleophilicity of the solvent. This equation is mainly of interest as a recognition of the complexity of solvent–solute interactions and the consequent inadequacy of any single measure of solvent polarity. The functions involved in the equation not being directly accessible, the quantitative application of the equation is not possible, but it was the basis of some qualitative discussion.

Swain, Mosely, and Bown (1955) put forward an analogous equation involving a nucleophilicity and an electrophilicity parameter of the solvent. Their equation was expressed in various forms. We content ourselves with quoting equation (83), where k^0 is the rate coefficient in the standard solvent,

$$\log(k/k^0) = c_1 d_1 + c_2 d_2 \tag{83}$$

c_1 and c_2 measure the sensitivity of the substrate to nucleophilic and electrophilic solvent character respectively, and d_1 and d_2 are measures of the nucleophilic and electrophilic character of the solvent. Scales were based on certain standard systems and numerous values of d_1, d_2, c_1, and c_2 were calculated, but the values are sometimes chemically absurd, e.g. t-butyl chloride has a higher c_1 value than methyl bromide, suggesting that methyl bromide (S_N2 mechanism) is *less* sensitive to solvent nucleophilicity than t-butyl chloride (S_N1 mechanism).

These and related attempts to allow for the complexity of solvent–solute interactions do not seem very fruitful. Presumably the parameters suggested fail to separate properly the factors involved.

Miscellaneous 'chemical' solvent parameters

Gielen and Nasielski have suggested that a polarity scale (X values) could be based on electrophilic aliphatic substitution, cf. Y and nucleophilic substitution. Doubtless X values are some measure of the ionizing power of the solvent, but again more specific effects probably play a role.

Among other chemical measures of solvent polarity which have been suggested are the $\log k$ values for the Menschutkin reaction of tri-n-propylamine and methyl iodide (20°C).

The Z-values

Kosower (1958) (see also Kosower, 1968) suggested that a polarity scale could be based on solvatochromic shifts. The charge-transfer absorption of 1-ethyl-4-methoxycarbonylpyridinium iodide shows marked solvatochromism, since the more polar solvents stabilize the ground state, which is an ion pair, relative to the excited state, which is a radical pair.

$$MeO_2C-C_5H_4\overset{+}{N}-Et \;\; I^- \xrightarrow{h\nu} MeO_2C-(7\pi)N-Et \;\; I^{\bullet} \tag{84}$$

The Z value for a solvent is defined as the energy of this transition (kcal mol^{-1}) at 25°C, a high Z-value corresponding to high solvent polarity. Z-Values cover the range 95 (water) to about 60 (iso-octane), and are available for a large number of one-component solvents and mixtures. Selected values are in Table 13. Z gives excellent correlations with other solvatochromic parameters and with certain rate data, e.g. Menschutkin reactions. For some mixed solvents Z is related linearly to Y.

E_T values of N-*phenol–pyridinium betaines*

The solvatochromic effect for the compound **31** with R = H is very large. The long-wave band is at 810 nm in diphenyl ether (transition energy in kcal mol^{-1}, E_T, of 35·3) and at 453 nm in water (E_T = 63·1). The more polar solvents stabilize the zwitter-ionic ground state relative to the excited state.

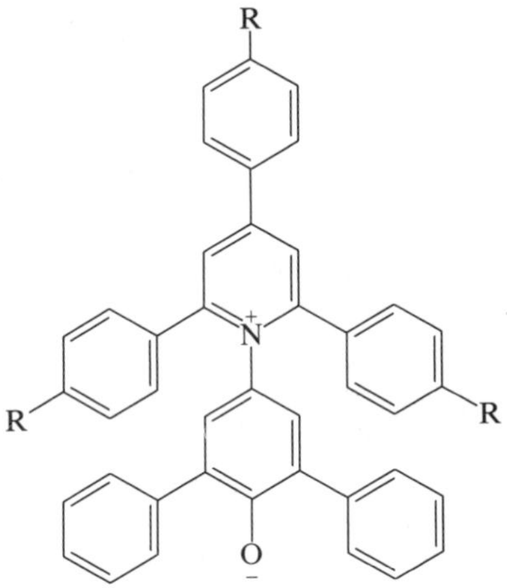

31

The compound with R = H is insoluble in hydrocarbons but the E_T scale can be extended to these by using the R = Me compound. Selected values of E_T at 25°C (from the work of Dimroth and his colleagues, see Reichardt, 1965) are in Table 13. Values are available for certain mixed solvents but not for acidic solvents which protonate **31** at the O^-.

Certain solvents, e.g. formamide and dimethyl sulphoxide (DMSO), which are commonly regarded as highly polar (they have high dielectric constants and they dissolve salts) are by no means so polar in their solvatochromic behaviour; the E_T values of DMSO and t-butyl alcohol are comparable. No doubt the ability of protic solvents to form hydrogen bonds with the O^- in **31** is important.

E_T-Values have been used for correlating various kinetic and spectroscopic data. Quite good correlations are sometimes observed, although often the range of solvents is restricted and there are 'badly-behaved' members.

Brownstein's LFER

By analogy with the Hammett equation Brownstein (1960) suggested (85) for the general description of solvent effects; k_S is the rate or equilibrium constant for a

$$\log(k_S/k_E) = SR \qquad (85)$$

reaction in various solvents, and k_E is the corresponding quantity for absolute ethanol as solvent. The equation also applies to solvatochromic effects, with $\log k$ replaced by the appropriate spectroscopic quantity. S is characteristic of the solvent ($S = 0{\cdot}00$ for ethanol) and R gives the susceptibility of the given property to change of solvent. The Kosower Z system was taken as standard ($R = 1{\cdot}00$). By application to a variety of processes, 158 S values and 78 R values were derived.

This is an interesting attempt at wide generalization but many of the correlations are rather poor. No doubt too many different physical effects are being mixed-up and treated in an oversimplified way.

Various other empirical measures of polarity based on spectra have been suggested; see the summary by Koppel and Palm (1972).

The detailed analysis of solvent effects

We have seen repeatedly that the concept of 'polarity' as a universally applicable solvent characteristic is a gross oversimplification. A number of different characteristics are embodied in it and they require separate specification. If this can be done the resultant parameters may be used to interpret solvent effects through multiple rather than simple correlation.

Koppel and Palm (1972) have suggested a division into *non-specific* and *specific* solvent–solute interactions. The former may be polarization or polarizability effects and are reasonably expressed by functions of dielectric constant or refractive index respectively. Specific interactions concern donor–acceptor interaction of solvent with solute. A solvent may function as a Lewis base (electron donor) capable of nucleophilic solvation, or as a Lewis acid (electron acceptor) capable of electrophilic solvation. Each solvent requires four parameters for complete specification of its behaviour, and correlation analysis should therefore involve equation (86). A is the value of the solvent-dependent property ($\log k$, ν, etc.) in a given solvent, and A_0 is the statistical

$$A = A_0 + yY + pP + eE + bB \qquad (86)$$

quantity corresponding to the value of the property in the gas phase as reference 'solvent'. Y†, P, E, and B are respectively parameters on scales of polarization, polarizability, Lewis acidity (electrophilic solvating power) and Lewis basicity, and y, p, e, and b are the corresponding regression coefficients. Such an equation can be legitimately applied only to data for a large number of well-chosen solvents, and its success must be examined by proper statistical methods (see Appendix). This imposes severe limits to its range of application at present.

Dielectric constants are the basis of Y, and are used either in the form of the Kirkwood function $(\varepsilon-1)/(2\varepsilon+1)$ or of a function based on the expression for molar polarization, $(\varepsilon-1)/(\varepsilon+2)$. Since these functions are linearly related, the choice is arbitrary. The corresponding functions $(n_D^2-1)/(2n_D^2+1)$ or $(n_D^2-1)/(n_D^2+2)$ are used for the polarizability function P. A scale for Lewis acidity, E, is based on the E_T-values discussed on p. 73, but these are corrected

† Do not confuse with the Grunwald–Winstein Y, see p. 70.

for the influence of non-specific effects, and adjusted to an origin $E = 0$ for the gas phase. (The basis for supposing that E_T largely relates to Lewis acidity is the exposed situation of the O^- in **31**. This is readily accessible to solvent, whereas the N^+ is buried in the molecule.) A scale for Lewis basicity, B, is based on the solvatochromism of ν_{OD} for CH_3OD. Electron-donating solvents reduce ν_{OD} through hydrogen bonding to the D atom,

$$CH_3—O—D\cdots\cdots:X—R$$

32

and the frequency shift measures the strength of the donor–acceptor interaction. The shifts are adjusted to an origin $B = 0$ for the gas phase.

Values of ε and n_D are readily available for many solvents; far fewer values of E and B are available, and the number of solvents for which both E and B are known is very restricted. Values of ε, n_D, E, and B for selected solvents are in Table 14. The results of analysing the solvent dependence of a few well-known processes are in Table 15, and will be discussed briefly.

TABLE 14

Selected values of ε, n_D, E, and B parameters†

Solvent	ε	n_D	E	B
(Gas phase	1·000	1·0000	0	0)
Dioxan	2·21	1·4224	3·98	129
Carbon tetrachloride	2·23	1·4603	(0)‡	31
Benzene	2·27	1·5011	1·93	52
Anisole	4·3	1·5170	(0)‡	78
Diethyl ether	4·3	1·3527	(0)‡	129
Chloroform	4·7	1·4180	3·17	35
Chlorobenzene	5·6	1·5218	(0)‡	50
Acetic acid	6·2	1·3716	14·48	98§
Aniline	6·9	1·5855	6·15	210
Tetrahydrofuran	7·4	1·4076	(0)‡	142
t-Butyl alcohol	12·2	1·3848	5·15	—
Pyridine	12·3	1·5100	(0)‡	220
i-Propyl alcohol	18·3	1·3773	8·70	—
Acetone	20·5	1·3588	2·13	116
Ethanol	24·3	1·3614	11·57	—
Methanol	32·7	1·3286	14·94	—
Nitrobenzene	34·8	1·5546	0·3	73
NN-dimethylformamide	36·7	1·4272	2·60	159
Methyl cyanide	37·5	1·3416	5·21	101
Dimethyl sulphoxide	46·6	1·4783	3·7	193
Water	78·5	1·3330	21·8	123§

† Data mainly from Koppel and Palm (1972). Values of ε at 25°C, n_D at 20°C.
‡ $E = 0$ presumed on structural grounds.
§ B cannot be measured directly for hydroxylic solvents. These values were based on malonic acid decomposition; see Table 15, p. 76.

TABLE 15

Detailed analysis of solvent effects†

Process	A_0	y	p	e	b	R	$s\%$‡	Solvents
$\log k$ (sec^{-1}), solvolysis of Bu^tCl, 25°C	−19·89	13·39	13·46	0·378	0·0	0·982	4·6	23—protic and aprotic
$\log k$ ($l\,mol^{-1}\,min^{-1}$), reaction of MeI with Pr^n_3N, 20°C	−9·00	9·18	13·81	0·345	0·0	0·945	8·6	29—aprotic
Z values ($kcal\,mol^{-1}$)	51·7	14·35§	0·0	1·309	0·0	0·968	6·6	27—protic and aprotic
$5 + \log k$ (sec^{-1}), decomposition of malonic acid, 120°C	−0·734	2·99	0·0	0·0	0·0112	0·992	4·4	11—aprotic and amine
$\log k$ ($l\,mol^{-1}\,min^{-1}$), reaction of benzoic acid and diazodiphenylmethane, 37°C	−2·14	4·06	6·98	0·182	−0·0189	0·980	7·3	16—aprotic
$\log k$ ($l\,mol^{-1}\,sec^{-1}$), reaction of aniline with benzoyl chloride, 25°C	−3·27	3·70	0·0	0·174	0·0164	0·903	16·3	12—aprotic

† From Koppel and Palm (1972). $Y = (\varepsilon-1)/(2\varepsilon+1)$ and $P = (n_D^2-1)/(n_D^2+2)$ except as indicated below.
‡ See Appendix, p. 107.
§ $Y = (\varepsilon-1)/(\varepsilon+2)$.

Rates of ionization of t-butyl chloride in a variety of one-component solvents are well correlated by Y, P, and E terms. The nucleophilic behaviour of the media is unimportant. The contribution of eE to the change in $\log k$ over the range from gas phase to water is easily the most important, and presumably indicates facilitation of the withdrawal of Cl^- by hydrogen bonding, etc.

The reaction between methyl iodide and tri-n-propylamine in aprotic solvents, shows behaviour similar to t-butyl chloride ionization but alcohols deviate markedly. Apparently these stabilize the amine reactant by hydrogen bonding.

Z values show good correlation with the Y and the E term, the P and the B term making no significant contribution. The positive sign of e indicates relative electrophilic stabilization of the initial state. The composite nature of Z is clearly shown, although over the range gas phase to water eE makes the larger contribution to the change in Z.

Nucleophilic stabilization of the activated complex is indicated for the decomposition of malonic acid. The reaction involves a dipolar activated complex **33** and basic solvents may presumably form helpful hydrogen bonds

$$\left[HO_2CCH_2C\overset{\delta-}{O}_2\cdots\overset{\delta+}{H}\right]^{\ddagger}$$

33

with the nascent hydrogen ion.

Certain systems show the simultaneous influence of electrophilic and nucleophilic solvation acting either to reinforce each other or in opposition. The reaction between carboxylic acids and diazodiphenylmethane in aprotic solvents gives significant correlation with all four parameters. b is negative and e is positive. This probably indicates nucleophilic stabilization of the carboxylic acid **34** and electrophilic stabilization of the activated complex **35**.

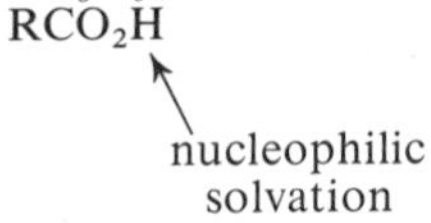

$$RC\overset{\delta-}{O}_2\overset{\delta+}{H}$$

(arrow to H: nucleophilic solvation)

34

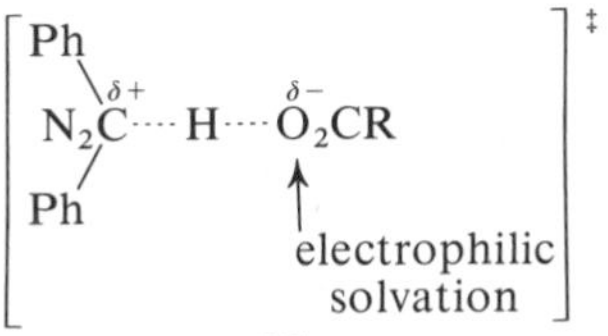

$$\left[Ph_2N_2\overset{\delta+}{C}\cdots H\cdots\overset{\delta-}{O}_2CR\right]^{\ddagger}$$

(arrow to O: electrophilic solvation)

35

The reaction of benzoyl chloride with aniline appears to be assisted by both types of solvation. Possibly the activated complex **36** requires nucleophilic

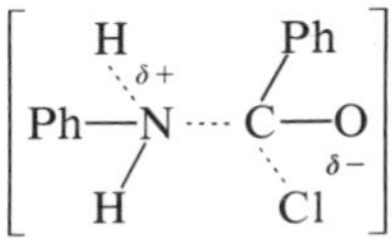

$$\left[Ph\text{—}\overset{\delta+}{N}H_2\cdots C(Ph)(\text{—}O)\cdots\overset{\delta-}{Cl}\right]^{\ddagger}$$

36

assistance for the removal of the proton and electrophilic assistance with the chloride ion.

Koppel and Palm (1972) give many other examples. While there is much that still remains obscure and unsatisfactory, their 'fundamental' approach to correlation analysis in the realm of solvent effects seems to offer great hope for the future in an area of physical organic chemistry where real understanding has been very slight.

PROBLEMS

5.1. (*a*) Show by long division that the Kirkwood function of dielectric constant may be expanded as the following series.

$$\frac{\varepsilon - 1}{2\varepsilon + 1} = \frac{1}{2} - \frac{3}{4\varepsilon} + \frac{3}{8\varepsilon^2} - \frac{3}{16\varepsilon^3} \text{ etc.}$$

Find the rough value of dielectric constant above which the approximation

$$\frac{\varepsilon - 1}{2\varepsilon + 1} = \frac{1}{2} - \frac{3}{4\varepsilon}$$

holds to within one percent.

(*b*) When the above approximation is valid, show that equation (74) may be rewritten as

$$\ln k = \ln k_\infty + \frac{3N_A}{4\varepsilon RT}\left[\frac{\mu_A^2}{r_A^3} + \frac{\mu_B^2}{r_B^3} - \frac{\mu_\ddagger^2}{r_\ddagger^3}\right]$$

where k_∞ is the rate constant for a medium of infinite dielectric constant.

(*c*) Show that equation (77) may be rewritten as

$$\ln k = \ln k_\infty + \frac{N_A Z^2 e^2}{2\varepsilon RT}\left(\frac{1}{r} - \frac{1}{r_\ddagger}\right)$$

5.2. The Table gives rate constants ($l\,mol^{-1}\,min^{-1}$) for the reaction of methyl iodide with tri-n-propylamine at 20°C. [The data are from Lassau, C. and Jungers, J.-C. (1968). *Bull. Soc. chim. France* 2678.]

Solvent	10^3k	*Solvent*	10^3k
Diethyl ether	1·2	Dioxan	37
Carbon tetrachloride	1·4	Bromobenzene	89
Carbon bisulphide	2·5	Chloroform	130
n-Propyl alcohol	7·4	Acetone	150
Ethanol	9·5	Dichloromethane	280
Toluene	9·5	Methyl cyanide	470
Methanol	13	DMF	600
Benzene	18	Nitromethane	1100

Is there any clear relationship of $\log k$ to E_T? (Values of E_T for some of the solvents are in Table 13. Others needed are as follows: Et_2O, 34·6; CS_2, 32·6; Pr^nOH, 50·7; PhMe, 33·9; PhH, 34·5; PhBr, 37·5; CH_2Cl_2, 41·1; $MeNO_2$, 46·3.) What is the significance of your findings?

6. The effect of the reagent on reactivity

Introduction

IN Chapters 2 and 3 we dealt with correlation analysis in relation to the substrate of an organic reaction, i.e. we were concerned with the influence of substituents on the reactivity of a given functional group undergoing reaction with a given *reagent*. Correlation analysis may also be applied to reaction series involving a given substrate reacting with various reagents. One aspect of this has already been examined in Chapter 5: solvent effects in solvolytic reactions, such as the S_N1 reactions of alkyl halides. As we shall see, the role of the solvent cannot be neglected in the wider aspects dealt with in the present chapter. Two main topics are usually distinguished: the influence of the catalyst in proton transfer reactions showing general acid or general base catalysis (the Brønsted equation) and the nucleophilicity of reagents in bimolecular reactions. These topics are not totally unconnected, however. We shall see that the application of correlation analysis in this field is still in a relatively elementary state of development and some qualitative discussion is therefore appropriate here.

Acid–base catalysis: the Brønsted equation

The catalysis of organic reactions by acids and bases is a very complicated subject and for detailed accounts larger texts must be consulted, e.g. Bell (1941, 1959) and many textbooks of physical organic chemistry (see Bibliography, p. 115). We shall develop the subject in a considerably simplified manner, as necessary background for the application of correlation analysis.

Some forms of acid-catalysis

The simplest situation is represented by the following scheme.

Type (*a*)
$$S + H^+ \overset{K}{\rightleftharpoons} SH^+ \tag{87}$$

$$SH^+ \xrightarrow{k_1} \text{products} \tag{88}$$

S is the substrate and H^+ is the hydrogen ion in a protic solvent, e.g. H_3O^+ in water. The scheme involves the equilibrium for rapid reversible protonation of the substrate, followed by the rate-limiting conversion of the protonated form into products. It is easily shown that the rate of reaction, v, is given by equation (89) and (90), where k_a is the catalytic constant. The hydrolysis of acetals,

$$v = k_1K[H^+][S] \tag{89}$$

$$k_a = k_1K \tag{90}$$

$RCH(OEt)_2$, shows this behaviour. Protonation leads to $[RCH(OEt)(HOEt)]^+$, which undergoes slow loss of ethanol to give the oxocarbonium ion $[RCH(OEt)]^+$; the latter undergoes a series of very rapid changes leading ultimately to the aldehyde, RCHO. This is often described as *specific* hydrogen ion catalysis.

A slightly more complicated scheme may apply when the substrate has an ionizable proton which is involved in the reaction.

Type (*b*)
$$\mathbf{HS} + H^+ \overset{K}{\rightleftharpoons} \mathbf{H}\overset{+}{S}H \tag{91}$$

$$\mathbf{H}\overset{+}{S}H + A^- \xrightarrow{k_1} H\mathrm{A} + SH \tag{92}$$

HS is the substrate, and A^- and HA are the components of a buffer solution controlling $[H^+]$. The heavy type is to distinguish the ionizable proton of the substrate from the hydrogen ions in solution. The HS is converted into an isomeric species SH, which can undergo reaction with other species present. In the ideal case such reactions will be very rapid, and the observed overall rate of reaction will be given by equations (93) to (96)

$$v = k_1K[HS][H^+][A^-] \tag{93}$$

or
$$v = k_1KK_a[HS][HA] \tag{94}$$

where
$$K_a = [H^+][A^-]/[HA] \tag{95}$$

$$k_a = k_1KK_a \tag{96}$$

and k_a is the catalytic constant for catalysis by the acid HA. Note that the actual processes involve H^+ and A^-.

In principle HA may be any (Brønsted) acid, so this scheme is for *general* acid catalysis.

Processes of the above type certainly play a part in the enolization of acetone, which may be studied by halogenation (Br_2 or I_2) in acidic buffer solutions, as indicated below.

$$CH_3COCH_3 + H^+ \underset{\text{fast}}{\overset{\text{very}}{\rightleftharpoons}} CH_3C(OH)^+CH_3 \tag{97}$$

$$CH_3C(OH)^+CH_3 + A^- \xrightarrow{\text{slow}} CH_3C(OH){:}CH_2 + HA \tag{98}$$

$$CH_3C(OH){:}CH_2 + X_2 \xrightarrow[\text{fast}]{\text{very}} CH_3COCH_2X + H^+ + X^- \tag{99}$$

It is also possible for HA to be involved directly in the attack on the substrate

Type (*c*)
$$S + HA \xrightarrow{k_1} SHA \tag{100}$$

$$SHA \xrightarrow{\text{fast}} \text{products} \tag{101}$$

The observed rate of reaction is that of the first (slow) step and is given by

(102), the catalytic constant k_a being k_1.

$$v = k_1[S][HA] \tag{102}$$

The acid-catalysed decomposition of diazodiphenylmethane in alcoholic solution is of this type:

$$Ph_2CN_2 + RCO_2H \xrightarrow{\text{slow}} \underset{\text{ion pair}}{[Ph_2\overset{+}{C}H{\cdot}N_2 \cdots \overset{-}{O}_2CR]} \tag{103}$$

$$[Ph_2\overset{+}{C}H{\cdot}N_2 \cdots \overset{-}{O}_2CR] \xrightarrow[\text{fast}]{\text{very}} Ph_2\overset{+}{C}H + N_2 + RCO_2^-. \tag{104}$$

The diphenylmethyl cation very rapidly reacts further either with RCO_2^- to give the diphenylmethyl ester of the acid or with the solvent alcohol R^1OH to give the ether, Ph_2CHOR^1.

The Brønsted equation

A survey of reactions showing general acid catalysis reveals that the catalytic constant, k_a, is related to the strength of the catalysing acid: the stronger the acid, the better its catalytic properties. Some qualitative connection between k_a and K_a, the dissociation constant of the acid, is quite reasonable in so far as k_a is concerned with ease of proton transfer to the substrate, and K_a measures extent of proton transfer to water.

$$HA + H_2O \rightleftharpoons A^- + H_3O^+ \tag{105}$$

For many reactions and acid catalysts, however, a precise quantitative relationship is well obeyed, and following Brønsted and Pedersen (1924), this relationship is usually written in the form

$$k_a = G_a K_a^\alpha \tag{106}$$

where G_a and α are constants characteristic of reaction, reaction medium, and temperature. Values of α commonly lie in the range 0·3–0·9. The Brønsted equation may be rewritten in the form

$$\log k_a = \log G_a + \alpha \log K_a \tag{107}$$

and is then obviously a linear free-energy relationship, of which it was probably the earliest example.

Certain reactions are catalysed specifically by hydroxide ions, but many show general base catalysis, see Bell (1941, 1959). For the latter the Brønsted equation takes the form

$$k_b = G_b K_b^\beta \tag{108}$$

where K_b is given by (110) for equilibrium (109).

$$B + H_2O \rightleftharpoons BH^+ + OH^- \tag{109}$$

$$K_b = [BH^+][OH^-]/[B] \tag{110}$$

These relationships were developed by Brønsted and Pedersen in their work on the base-catalysed decomposition of nitramide

$$NH_2NO_2 \longrightarrow N_2O + H_2O. \tag{111}$$

In either of its forms the Brønsted equation has been successfully applied to many reactions. It is fairly obvious why type (*c*) processes should conform to the Brønsted equation: HA attacks the substrate directly. It is less obvious for type (*b*), for which $k_a = k_1KK_a$ and k_1 is for the attack of $H\overset{+}{S}H$ by A^-. Conformity to the Brønsted equation implies that

$$k_1KK_a = G_aK_a^\alpha \tag{112}$$

Since K and G_a are characteristic of the reaction (substrate), solvent, and temperature, it follows that k_1 is proportional to $K_a^{\alpha-1}$. Since $\alpha < 1$, this means that k_1 must decrease as K_a increases, i.e. the stronger the acid, the poorer is the anion at removing a proton from $H\overset{+}{S}H$, which is rational.

The inclusion in Brønsted correlations of acid catalysts with more than one acidic centre requires a *statistical correction*. Consider an acid with two equivalent acidic centres, e.g. succinic acid. There is a statistical factor of two assisting both the catalytic action and the ionization of the acid. If such an acid is to be fitted into a Brønsted correlation along with catalysts having only one acidic centre, k_a and K_a must be reckoned *per acidic centre*, i.e. $k_a/2$ and $K_a/2$ must be used for the two-centre catalyst. A related problem arises for the monoanion of the dibasic acid. There are more difficult problems when the acidic centres are not equivalent, and for multiple attachment of dissociable protons to the same atom. The statistical effect also arises in base catalysis, see Bell (1941, 1959).

We now examine a few of the systems to which the Brønsted equation has been applied. One of the best explored reactions is the dehydration of acetaldehyde hydrate in aqueous acetone (Bell and Higginson, 1949). Some of the results are plotted in Fig. 10. Various carboxylic acids, phenols, and NH-acids as catalysts conform well to a linear relationship over ten units of pK_a and five orders of magnitude in k_a. Catalysts of other types, such as oximes, β-diketones, and nitroparaffins show marked deviations, some above and some below the line. The best Brønsted correlations usually involve catalysts of the same structural type. Deviations may be profitably discussed in terms of electronic and steric effects.

The decomposition of nitramide still provides one of the best examples of Brønsted correlation for a base catalysed reaction, see Bell (1941). For neutral bases or those with single or double negative charges a good linear relationship holds, although careful inspection shows that separate lines could legitimately be drawn for each class of base. A very different line is, however, followed by metallic complexes which function as bases. Certain tertiary amines also show marked deviations from the line based on nuclear substituted anilines. The

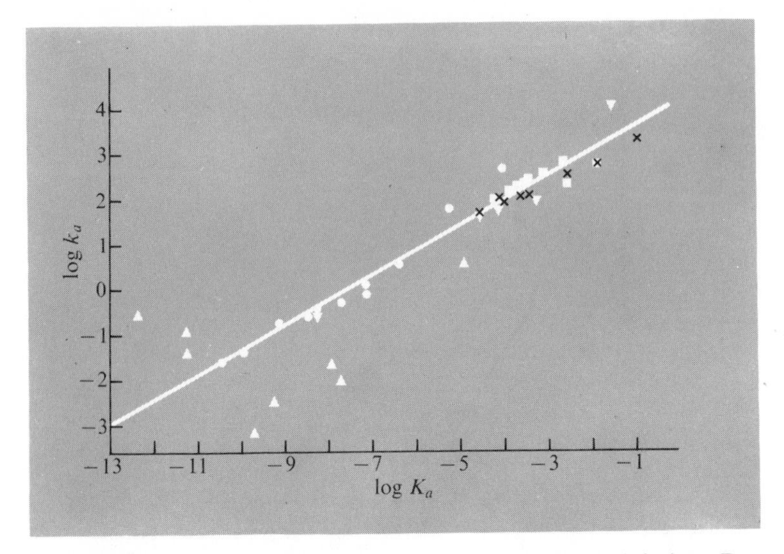

FIG. 10. Brønsted plot for acid-catalysed dehydration of acetaldehyde hydrate. Data from Bell and Higginson (1949); k_a in l mol^{-1} min^{-1}, 25°C. Statistical corrections to data have been applied where necessary. Catalysts as follows:
○ phenols; × aliphatic carboxylic acids; □ benzoic acids; ▽ NH acids; △ miscellaneous acids, including oximes and aliphatic nitrocompounds.

tertiary amines are much more effective catalysts than corresponds to their pK_b values.

From values of α or β the extent of proton transfer in the activated complex is often inferred. In such cases K_a and k_a (or K_b and k_b) should relate to the same solvent and temperature. For the reaction between diazodiphenylmethane and carboxylic acids in ethanol, $\alpha \sim 0{\cdot}65$. This may be taken to mean that the proton is rather more than half transferred from the carboxy group to the central carbon atom of the diazodiphenylmethane in the activated complex.

In recent years the development of techniques for studying very fast reactions has greatly extended our knowledge of proton transfer processes, and has exposed a general limitation of Brønsted correlations. It seems likely that all Brønsted plots would be revealed as curves if catalysts with a sufficiently wide range of pK_a or pK_b values could be studied.

The process

$$[CH_3{\cdot}CO{\cdot}CH{\cdot}CO{\cdot}CH_3]^- + HA \underset{k_2}{\overset{k_1}{\rightleftharpoons}} CH_3{\cdot}CO{\cdot}CH_2{\cdot}CO{\cdot}CH_3 + A^- \quad (113)$$

has been studied for catalysts covering about seventeen units of pK_a. The relationship between $\log k_1$ and $\mathrm{p}K_a$ is shown in Fig. 11. For phenols and

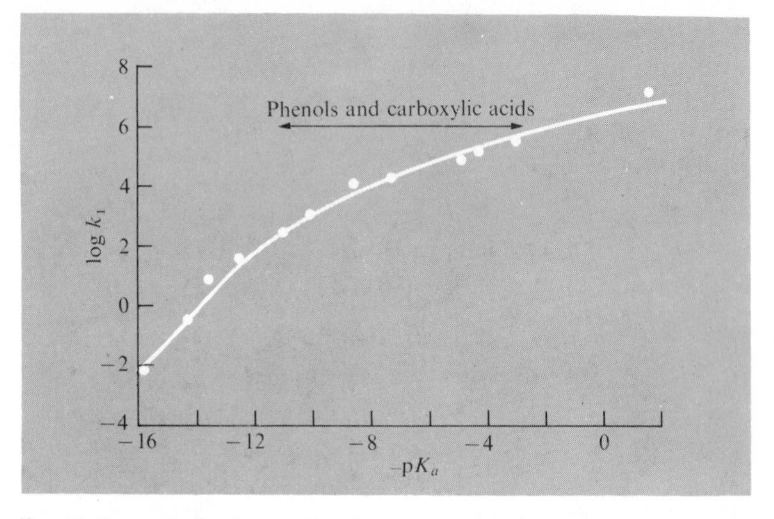

FIG. 11. Brønsted plot for transfer of proton to enol of acetylacetone. Data from Eigen (1964). k_1 in l mol^{-1} sec^{-1}, 12°C.

carboxylic acids as catalysts a straight line could be drawn with $\alpha \sim 0{\cdot}4$ but the extension of the catalyst range reveals pronounced curvature, with α tending to 1·0 at one end and to zero at the other. Detailed explanation of this and similar behaviour is quite complicated. However, the flattening at one end ($\alpha = 0$) is connected with the tendency of the process to become diffusion-controlled when HA is a very strong acid. For extremely weak acids the reverse reaction tends towards the diffusion-controlled rate, and k_1 becomes proportional to the equilibrium constant for the reaction and hence to the ordinary dissociation constant of the acid HA ($\alpha = 1{\cdot}0$).

Nucleophilicity in bimolecular reactions

A nucleophilic displacement reaction may be regarded essentially as an acid–base reaction; the nucleophile (base, electron-donor) attacks the substrate (Lewis acid, electron-acceptor). In view of the success of the Brønsted relationship in proton transfer processes, some relationship might be expected between the ability of an anion A^- to act as a nucleophile in displacement reactions and the $\mathrm{p}K_a$ value of the acid HA; the best nucleophiles should be the anions of the weakest acids. For the neutral nucleophile there should be a relationship with the $\mathrm{p}K_b$ value or the $\mathrm{p}K_a$ value of BH^+.

There are indeed a number of examples in which $\log k$ for a nucleophilic displacement reaction may be correlated with a pK value, e.g. $\log k$ values for various substituted anilines reacting with benzoyl chloride in benzene solution are linearly related to pK_a values for the anilinium ions in aqueous solution. The linear relationship includes *ortho*-substituted anilines.

When the reactions studied involve a wide range of nucleophiles the linear relationship is less clear, as in Fig. 12 for the hydrolysis of *p*-nitrophenyl

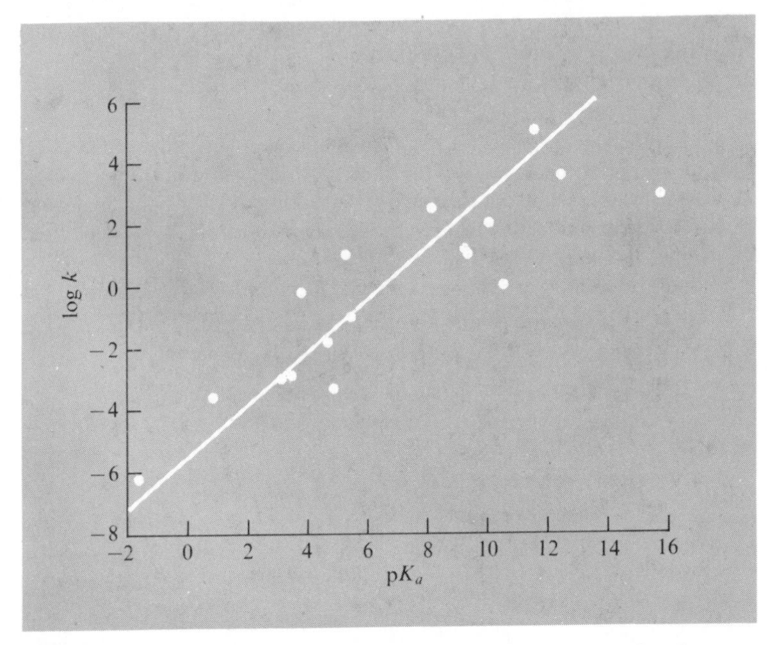

FIG. 12. Attempted Brønsted plot for nucleophilic attack on *p*-nitrophenyl acetate. After Jencks and Carriulo (1960). k in l mol^{-1} min^{-1}, 25°C; pK_a values are for the conjugate acids; no statistical corrections have been applied, but plot would not be much changed if they were. Nucleophiles were a variety of inorganic and organic bases, involving various donor atoms; the species were variously charged (±) or neutral.

acetate initiated by nucleophilic attack on the carbonyl group of the ester.

$$p\text{-NO}_2\text{C}_6\text{H}_4\text{OCOCH}_3 + \text{B} \xrightarrow{\text{slow}} \text{CH}_3\text{COB}^+ + p\text{-NO}_2\text{C}_6\text{H}_4\text{O}^- \qquad (114)$$

$$\text{CH}_3\text{COB}^+ + \text{H}_2\text{O} \xrightarrow{\text{fast}} \text{CH}_3\text{CO}_2\text{H} + \text{BH}^+ \qquad (115)$$

We have already noted the same kind of scatter when a great variety of acidic or basic catalysts is used in an ordinary Brønsted plot for proton-transfer processes. Further, it is easy to find nucleophilic displacement reactions for which any connection between nucleophilicity and pK is tenuous in the extreme. Such a situation is presented by the attack of nucleophiles on methyl iodide in methanol; a plot of $\log k$ versus pK_a-values for about thirty nucleophiles shows an almost random distribution of points.

This situation suggests that the development of correlation analysis for nucleophilic displacement reactions should concentrate initially on operating within the system of such reactions alone, without reference to the basicity of nucleophiles as measured by pK_a values. Swain and Scott (1953) suggested a linear free-energy relationship for nucleophilicity,

$$\log(k/k^0) = sn \tag{116}$$

where k is the rate constant for reaction of a given nucleophile with a given substrate, k^0 is for the reaction of the same substrate with water, n is a nucleophilicity constant (analogous to σ), and s is a susceptibility constant (analogous to ρ). Methyl bromide was taken as the standard substrate (s = 1·00, 25°C); n for water was set equal to 0·00, at 25°C.

Selected values of n are shown in Table 16, and are discussed below. The

TABLE 16

Parameters for nucleophiles†

Nucleophile	pK_a‡ (H−1·74)	n	n_{Pt}	E_n
ClO_4^-	—	<0·00	—	—
H_2O	−1·74	0·00	—	0·00
MeOH	—	—	0·00	—
NO_3^-	(−1·3)	1·03	—	(0·29)
F^-	3·45	2·00	<2·2	−0·27
SO_4^{2-}	2·0	2·50	—	0·59
OAc^-	4·8	2·72	<2·0	(0·95)
Cl^-	(−4·7)	3·04	3·04	1·24
C_5H_5N	5·23	3·6	3·19	(1·20)
HPO_4^{2-}	7·2	3·8	—	—
Br^-	(−7·7)	3·89	4·18	1·51
N_3^-	4·74	4·00	3·58	(1·58)
$CS(NH_2)_2$	−0·96	4·1	7·17	2·18
OH^-	15·7	4·20	—	1·65
OMe^-	—	—	<2·4	—
$PhNH_2$	4·6	4·49	3·16	(1·78)
SCN^-	(−0·7)	4·77	5·75	1·83
I^-	(−10·7)	5·04	5·46	2·06
CN^-	9·3	5·1	7·14	2·79
$S_2O_3^{2-}$	1·9	6·36	7·34	2·52

† pK_a, n_{Pt}, and E_n values mainly from the compilation by Pearson (1972). n values mainly from Swain and Scott (1953). Values in parentheses estimated indirectly and are uncertain.

‡ pK_a of the conjugate acid.

Swain–Scott relationship may be used to correlate the results of a variety of bimolecular nucleophilic displacement reactions, as shown in Fig. 13.

There is relatively little connection between n and pK_a-values overall, but a

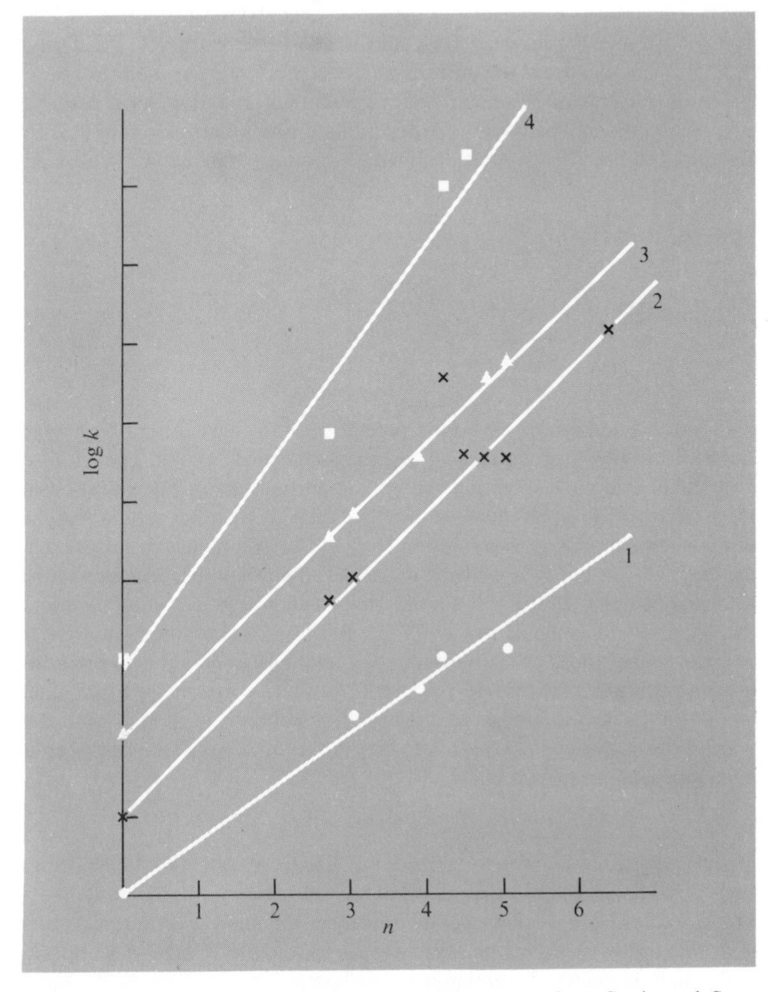

FIG. 13. Application of the Swain–Scott equation. Data from Swain and Scott (1953). The divisions on the ordinate are 1·00 unit of log k apart, and the relative positions of the lines 1 to 4 with respect to the ordinate are arbitrary. The substrates are: 1. ethyl tosylate; 2. mustard cation; 3. epichlorohydrin; 4. benzoyl chloride. See the original paper for further details. Nucleophiles may be identified from n-values in Table 16.

connection can be seen within limited ranges of nucleophiles. Thus the anions of the strongest oxyacids are very poor nucleophiles, and the order $ClO_4^- < NO_3^- < SO_4^{2-} < OAc^- < HPO_4^{2-} < OH^-$ seems reasonable in relation to pK_a-values. However, the halide ions show the order $I^- > Br^- > Cl^- > F^-$; thus I^- (derived from the strong acid HI) is a better nucleophile than F^- (from the weak acid HF), and is also better than OH^-. Different structural factors appear to control n-values and pK_a-values.

One obvious factor is ionic radius: large anions tend to be more reactive than small anions. Presumably large anions, being polarizable, are able to minimize electrostatic repulsion in the trigonal bipyramidal activated complex **37**.

$$\left[\overset{\delta_1-}{X}\cdots C \cdots \overset{\delta_2-}{Br}\right]^{\ddagger}$$

37

Polyatomic anions are especially potent, since the partial negative charge can be distributed over various atoms, e.g. with the SCN^- ion. There is also little doubt that ionic solvation is a very important factor. The solvation of anions in hydroxylic solvents was described on p. 69. It seems certain that the order of n-values in aqueous solution $I^- > Br^- > Cl^-$ is due to solvation in the order $Cl^- > Br^- > I^-$ having to be broken down in the formation of the activated complex. In the absence of such solvation, e.g. in acetone solution, the order of nucleophilicities is $Cl^- > Br^- > I^-$. Thus the Swain–Scott equation as developed above is essentially a relationship for aqueous solutions or aqueous organic mixtures (n-values will, however, correlate the $\log k$ values for MeI in methanol referred to already). It is not difficult to find substrates which present nucleophilic reactivity orders very different from the order of n-values. Thus the reaction:

$$B + trans\text{-}Ptpy_2Cl_2 \longrightarrow trans\text{-}Ptpy_2ClB^+ + Cl^- \quad (117)$$

has been suggested as another reference reaction for assessing nucleophilicities (py = pyridyl). Selected n_{Pt}-values are also shown in Table 16. While the n and n_{Pt} scales show a rough parallelism there are many discrepancies. It is clear that the Swain–Scott equation is much too simple to be of wide application, whatever standard reaction may be chosen.

Edwards (1954) suggested an equation to express the operation of two factors governing nucleophilicity. The equation takes the form:

$$\log(k/k^0) = \alpha E_n + \beta H \quad (118)$$

where α and β are characteristic of the substrate and E_n and H of the nucleophile. The quantity E_n is related to the electrode potential for a redox half-reaction in which the nucleophile is the reduced form, e.g.

$$2I^- \rightleftharpoons I_2 + 2e^-. \quad (119)$$

The actual definition of E_n is

$$E_n = E^0 + 2{\cdot}60 \quad (120)$$

where E^0 is the electrode potential for the half-reaction and the 2·60 is introduced to give for water $E_n = 0{\cdot}00$, with E^0 for the half-reaction

$$2H_2O \rightleftharpoons H_4O_2^{2+} + 2e^- \quad (121)$$

being −2·60 volts.

H is defined by the equation

$$H = pK_a + 1{\cdot}74 \quad (122)$$

where pK_a is for the conjugate acid of the nucleophile and 1·74 comes from the notional pK_a-value for H_3O^+, −1·74. Selected values of E_n and H are in Table 16.

The Edwards equation is essentially a Brønsted relationship (the βH term) modified by a term (αE_n) which expresses the polarizability of the nucleophile. E_n is clearly related to the ease of removal of electrons from the nucleophile, which may in turn be held to reflect its polarizability. A large value of β/α indicates a Brønsted-type behaviour, while a large value of α/β indicates the dominant role of nucleophile polarizability.

The equation is quite successful for a variety of displacement reactions but its application is limited by the relatively small number of nucleophiles for which electrochemically-determined values of E_n-exist. Other values have been determined by interpolation on Edwards plots. Further, the physical significance of the two terms in the equation is doubtful since presumably polarizability influences H values as well as E_n. Since it seems certain that the hydration of anions is an important factor governing their nucleophilicity in water, it must be oversimplified to interpret failure of $\log(k/k^0)$ to be correlated with pK_a solely in terms of polarizability. Presumably E_n itself depends on both polarizability and hydration effects.

Hard and soft acids and bases

The difficulties of understanding nucleophilicity in quantitative terms through correlation analysis are considerable. However, a qualitative rationalization of the situation is possible through the application of the principle of hard and soft acids and bases (HSAB). For a detailed exposition of HSAB, the articles of R. G. Pearson should be consulted; he is the originator of the ideas [see Pearson (1968, 1972)].

Hard bases are those which hold on to their electrons very strongly, are not easily oxidized, and are of low polarizability. They mainly involve F, O, or N as the donor atom. Soft bases have the opposite characteristics and involve donor atoms such as P, S, and I. Hard acids are Lewis acids in which the acceptor atom has one or more of the following characteristics: it is small, of high positive charge, or lacking unshared pairs of electrons in the valence shell. Electrophiles such as H^+, Al^{3+}, BF_3, and RCO^+ are hard, whereas CH_3^+, $C_6H_5^+$, I^+, and Pt^{2+} are soft. A fundamental principle of HSAB is that hard bases react preferentially with hard acids and that soft bases do so with soft acids. Mismatched processes, soft with hard, occur much less readily.

According to Pearson the parallelism expressed in Brønsted relationships is connected with proton transfer reactions and acid-base equilibria both involving reactions of nucleophiles with the hard electrophilic centre, the proton. $\log k$ values for reactions at other electrophilic centres will show correlations with pK_a-values provided that the electrophilic centres are hard. This is more or less true for reactions at carbonyl groups (see p. 85). When the electrophilic centre is soft, as for the carbon atom of methyl iodide or the Pt^{II} atom of *trans*-$Ptpy_2Cl_2$, different factors influence reactivity, and correlation with pK_a values is far worse.

These ideas are developed into a very satisfying rationalization of a wide variety of data. For example, the order of softness for the halide ions is $I^- > Br^- > Cl^- > F^-$. In reaction with soft electrophilic centres the order of reactivity tends to be $I^- > Br^- > Cl^- > F^-$, while with harder centres the opposite order may be found. With very hard centres the larger ions, all of which are fairly soft, may not react at all, although the harder F^- ion may react.

The physical basis of rationalization in terms of HSAB is not always clear, and HSAB seems to be a convenient way of dealing with a variety of factors. The quantitative formulation of the ideas is as yet largely undeveloped, and will probably require much new experimental work, particularly involving non-aqueous solutions. This is another area where undue emphasis on aqueous solutions has retarded deeper understanding.

PROBLEMS

6.1. In Fig. 10 the three deviant points above the line at the left of the graph are for two oximes and chloral hydrate; the four well below the line at the left are for three aliphatic nitro-compounds and benzoylacetone. By considering the structures of these acids and their ions, suggest an explanation of the deviations.

6.2. The Table gives data for the nucleophile-catalysed hydrolysis of an arylsulphinyl sulphone in 60% dioxan at 21·4°C.

$$ArSO{\cdot}SO_2Ar + H_2O \xrightarrow{N} 2ArSO_2H$$

Nucleophile	k ($l\,mol^{-1}\,sec^{-1}$)
F^-	4·4
OAc^-	9·0
Cl^-	12·0
Br^-	65
SCN^-	$1{\cdot}7 \times 10^2$
I^-	$1{\cdot}0 \times 10^3$
$CS(NH_2)_2$	$3{\cdot}5 \times 10^3$

Construct a table showing $\log k$ and the appropriate values of pK_a, n, and n_{Pt} from Table 16. Draw and interpret whatever graphs seem useful.

7. Some biological effects of organic compounds†

Introduction

WE have seen how correlation analysis helps in understanding the influence of structure on reactivity in organic chemistry. We now look briefly at the interaction of organic compounds with biological material.

The most obvious field to examine is enzymology; it is closely related to 'ordinary' organic chemistry. We may hope to find that LFER apply to the influence of substrate or inhibitor structure on enzymic reactions.‡ Less closely related to organic chemistry is pharmacology. There is often a long chain of varied events between the introduction of a drug into a biological system and the observed biological response. Nevertheless drug activity is clearly subject to structural factors and again we may hope that LFER parameters find some application, albeit highly empirical.

Enzymology

We first examine briefly necessary background material. [For thorough treatments see Laidler (1958) and Dixon and Webb (1964).]

Enzymes are highly specific biological catalysts. Each enzyme catalyses only a limited number of reactions, sometimes only a single reaction of very few substrates. 'Proteolytic' enzymes catalyse the hydrolysis of peptide lingages (CO—NH) derived from L-amino acids. Other 'digestive' enzymes catalyse the hydrolysis of polysaccharides or esters of fatty acids. The name of an enzyme or class of enzymes is commonly based on its usual biochemical function rather than structure, about which little may be known. The source from which the enzyme may be isolated is often included in a full description, e.g. fly-head acetylcholinesterase (see p. 101).

Enzymes are essentially proteins. Catalytic activity is associated with a small region of the macromolecule, known as the *active centre*. The high specificity is due to the substrate molecule having to fit the active centre in a fairly precise way. Certain other substances are sometimes essential to enzymic action; these are *co-enzymes*.

Enzymic reactions are usually of the first order in enzyme, and first order in substrate at low substrate concentration, but tend to zero order in substrate at high concentration. The simplest scheme to explain this behaviour (Michaelis

† The use of biological technical terms in this Chapter has been minimized. The more important terms are defined, unless they have meanings which are easily understood in context. Unfamiliarity with, for example, bovine serum albumin, guinea pig ileum, or *Escherichia coli* should not prevent the reader from gaining a general appreciation of the application of correlation analysis to the biological effects of organic compounds.

‡ The compound on which the enzyme acts is known as the *substrate*. The action may be subject to interference by certain bodies, to which the term *inhibitor* is applied.

and Menten, 1913) is in (123) and (124). E, S, and P are respectively the enzyme,

$$E + S \underset{k_{-1}}{\overset{k_1}{\rightleftharpoons}} ES \quad (123)$$

$$ES \xrightarrow{k_2} E + P \quad (124)$$

substrate, and product, and ES is an intermediate complex. The latter is assumed rapidly to reach a steady concentration, and the usual type of kinetic treatment leads to expression (125) for the rate of reaction. $[E]_0$ is the total concentration of enzyme, i.e. free as E and combined as ES; K_m is $(k_{-1}+k_2)/k_1$

$$v = \frac{k_2[E]_0[S]}{K_m + [S]} \quad (125)$$

and is the *Michaelis constant*. When $[S] \ll K_m$,

$$v = \frac{k_2[E]_0[S]}{K_m} \quad (126)$$

and when $[S] \gg K_m$,

$$v = k_2[E]_0. \quad (127)$$

In (127) v may be denoted as v_{max}, and (125) may be written as

$$v = \frac{v_{max}[S]}{K_m + [S]} \quad (128)$$

v attains the value $v_{max}/2$ when $[S] = K_m$.

The rate of an enzymic reaction is commonly influenced by the pH of the medium, for the active centre often exists in various states of ionization, which can interact characteristically with the substrate. The resulting kinetics and interpretation can be very complicated.

LFER in enzymic reactions

In even the simplest enzymic reaction there are at least two steps: the formation of the complex and its breakdown. The situation is thus complicated for devising correlation equations. The most satisfactory procedure is to analyse the overall process into the component parts and to deal with these individually.

The binding of substrates to enzymes is one facet of the binding of small molecules to proteins. This does not necessarily involve the making or breaking of covalent bonds and the structural effects may not be closely related to those of organic chemistry. We treat this topic more specifically later in connection with drug action (see p. 98). Note here that, for the simple case, if $k_{-1} \gg k_2$, K_m becomes k_{-1}/k_1, and is denoted by K_s, the dissociation constant of the enzyme–substrate complex.

The more chemical part of the enzymic process, in the simple case, may be isolated by dealing with rates measured in the zero order region: $v_{max} = k_2[E]_0$

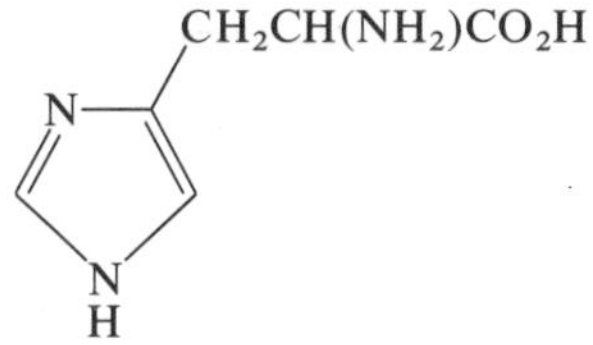

$CH_2CH(NH_2)CO_2H$

N

N
H

40

41. The structure may then be represented as **38** in equation (130). (N.B. This

$$HOCH_2CH(NH_2)CO_2H$$

41

mechanism is not typical of chymotrypsin action in general.)

The ability of chymotrypsin to accept a variety of substrates has great possibilities for LFER.

As for ordinary organic reactions, peculiar forms of Hammett plot may be obtained for enzymic reactions when competing or consecutive processes are involved (p. 19). The Hammett equation may also be of interest in connection with substituent effects on enzyme inhibition, see Kirsch (1972).

Further applications of correlation analysis

Studies of the deacylation of aliphatic acylchymotrypsins, and of the acylation of chymotrypsin by *p*-nitrophenyl esters of aliphatic carboxylic acids (straight or branched alkyl groups) show an important role for steric effects. The Pavelich–Taft equation (p. 41) involving σ^* and E_s may be applied to the specific rates of both reactions. The E_s-term makes the major contribution, with increasing bulk of the alkyl group retarding the reaction. In both reactions the n-hexyl compound is anomalously reactive, indicating perhaps the limitation of applying E_s-values to a system so different from that used to define E_s.

Acetylcholinesterase, which is important in the functioning of nerves, catalyses the hydrolysis of acetylcholine to acetic acid and choline, (131).

$$MeCO_2[CH_2]_2\overset{+}{N}Me_3 + H_2O \longrightarrow MeCO_2H + HO[CH_2]_2\overset{+}{N}Me_3 \qquad (131)$$

Correlation analysis has been widely employed in connection with the binding of inhibitors to acetylcholinesterase.

The inhibition of acetylcholinesterase by eight 2,4,5-trichlorophenyl methyl *N*-alkylphosphoramidates, **42**, is described fairly well

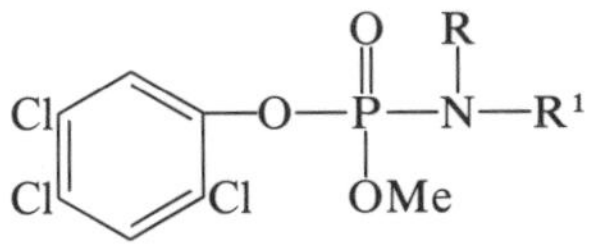

O R

Cl —O—P—N—R[1]

Cl Cl OMe

42

($R = 0{\cdot}94$) by equation (132). K_i is the inhibition constant, and relates to a

$$\log K_i = -4{\cdot}95 + 3{\cdot}91\sigma^* - 2{\cdot}36E_s \qquad (132)$$

process (133) whereby the active sites are blocked by the inhibitor I. Modification of the kinetic treatment already given introduces a new term involving

$$\mathrm{E} + \mathrm{I} \underset{k_{-i}}{\overset{k_i}{\rightleftharpoons}} \mathrm{EI} \qquad (133)$$

K_i, defined as k_{-i}/k_i. The E_s and σ^* terms in (132) actually involve summations for R and R^1, but since R is H in seven of the compounds and Me in the eighth, doubts about the validity of summing E_s-terms (see p. 42) are not important. Steric effects tend to be more important than polar effects, but electron-releasing substituents definitely promote inhibitory action (decrease K_i).

The active centres of acetylcholinesterase are anionic in nature (probably $-CO_2^-$) and these bind the positive nitrogen atom of the substrate. Alkylammonium ions act as inhibitors by competing for the active centres, but electrostatic forces are not the only consideration. Substantial effects of varying the alkyl groups are observed, e.g. plots of pI_{50} values (see p. 97 for definition), at a given substrate concentration, for n-alkyltrimethylammonium ions against n (the number of carbon atoms in the n-alkyl group) are rectilinear. The slope indicates an increment in binding energy of 300 cal per methylene group per mole.

The article by Kirsch (1972) gives many other examples of LFER in enzymology.

Pharmacology

Measures of biological activity

When a biological action of a compound is measured as a function of dose, a sigmoid relationship between response and dose is commonly observed. The limiting response at high dose is the basis for expressing responses on a percentage scale, and dose is conveniently expressed logarithmically to give the form of curve shown in Fig. 15.

The curves are for two compounds, that for the less potent being displaced to the right. Complete dose–response curves are costly to determine, and pharmacologists are sometimes content to measure varying biological responses towards a fixed dose in a particular series of compounds. This is useful for medical practice but not for correlation analysis: the fixed dose may be too low to elicit any response from the less active members and too high for the more active members, thus leading to an apparently constant response.

A more fundamental procedure is to express biological activity in terms of the dose required to produce a standard response, usually 50 per cent of the

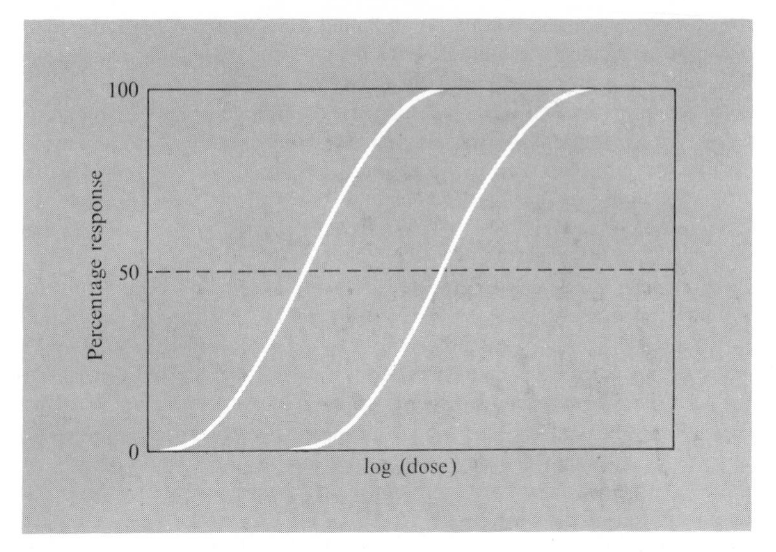

FIG. 15. Percentage response versus log (dose) curves. After Cammarata and Rogers (1972).

limiting response. Quantities such as the following are therefore used:

LD_{50} — the mean lethal dose, i.e. that required to kill 50 per cent of the tested population;

I_{50} — molar concentration of inhibitor required to halve the rate of a specified biological process;

ED_{50} — the mean effective dose of antagonist, i.e. that required to reduce by 50 per cent the response to a standard dose of the agonist.†

These quantities in a sense correspond to the Michaelis constant, K_m, in enzymology, i.e. the value of [S] to give a rate $v_{max}/2$. For correlation analysis we use these quantities in negative logarithmic form. Symbols such as pI_{50} may be used, but to avoid a multiplicity of symbols we shall normally use the general symbol $\log(1/C)$; all the measures of biological activity are effectively concentrations, in suitable molar units.

Biological activity may sometimes be expressed in terms of rate or equilibrium constants, e.g. the action of a drug on a growing bacterial population may be expressed in terms of growth constants, as in (134). k^0 is the first-order growth constant in the absence of the drug, k the diminished constant in the

$$k = k^0 - k_i c \tag{134}$$

† e.g. adrenaline (the agonist) raises blood pressure; its effect is counteracted, for example, by *NN*-dimethyl-2-bromophenylethylamine (the antagonist), see p. 101.

presence of the drug in concentration c, and k_i is the second-order growth inhibition constant for the drug. When drug action amounts to enzyme inhibition, K_i is appropriately used (see p. 96).

Microbiologists often use a measure MIC—minimum inhibitory concentration. This is a type of variable dose-fixed response quantity, but has a more complex meaning than the others mentioned already. It has found slight use in correlation analysis.

The complexity of biological processes

A biological response is the outcome of a complex series of events. The drug is administered in an aqueous phase outside the biophase and is then transported through various fluids and membranes in the biological material to the site of action. It becomes bound to the appropriate receptor and perturbs it. The perturbation issues in a biological response. In some studies (*in vitro*) the response is in an organ excised from its proper environment, e.g. muscular or nerve tissue. Alternatively a living organism may be used (*in vivo* study). In all cases, particularly *in vivo*, the biological response is separated from the administration of the drug by many and varied processes. When we apply correlation analysis, and particularly if we use LFER parameters from organic chemistry, we assume hopefully that somewhere along the line there will be a process (or processes) whose free-energy change governs the biological response and which may depend on electronic, steric, and other factors similar to those operating in inanimate matter. Remarkably this hope often appears to be justified.

As chemists we might rush into examining the applicability of σ, σ^*, E_s, etc. We must recognize, however, that pharmacologists have long realised that processes of membrane penetration and transport are often of dominant importance, i.e. the passage of the drug from the aqueous phase into the macromolecular material of the biophase. The structural factors influencing this are not those usual to organic chemistry but involve the *hydrophobic–lipophilic* character of molecules.

Hydrophobic–lipophilic influences on biological activity

What is meant by 'hydrophobic–lipophilic character'? The hydrocarbon part of a large organic molecule has little affinity for water molecules, but in aqueous solution it causes an increase in ordering in the sheath of water molecules surrounding it. The transfer of such a solute from aqueous solution to an organic solvent involves a decrease in free energy, mainly due to the entropy increase accompanying the de-structuring of the water sheath. Given the opportunity, the solute will forsake the aqueous solution for the organic solvent; the solute is *hydrophobic*. The same considerations apply to the tendency of organic molecules to leave aqueous solution for membrane material, proteins, etc.; the term *lipophilic* is equally appropriate.

The term *hydrophobic bonds* is sometimes used, but in a sense such 'bonds' are negative in character. In the main they owe their apparent strength to the tendency of water molecules to form strong hydrogen bonds with each other, and exclude molecules for which they have no particular affinity. There may be some inherent forces of attraction between organic molecule and, for example, protein; dipole–dipole, van der Waals, and hydrogen-bonding forces are sometimes important.

The partition of an organic solute between an organic solvent and water provides a possible measure of hydrophobic–lipophilic character. The partition coefficient for distribution between n-octanol and water is most commonly used (Hansch, 1969). This is denoted by P, and, as an equilibrium constant, is used in correlation analysis as $\log P$. For aromatic systems, a substituent index π is defined by (135), where P_X is the partition coefficient for a compound

$$\pi = \log P_X - \log P_H \qquad (135)$$

with X as substituent, and P_H is that for the parent compound. For various reasons the most commonly used π-scale is based on substituted phenoxyacetic acids. Selected values are in Table 17. P and π characterize the whole molecule,

TABLE 17

Selected values of π†

Substituent	π_o	π_m	π_p
F	0·01	0·13	0·15
Cl	0·59	0·76	0·70
Br	0·75	0·94	1·02
I	0·92	1·15	1·26
Me	0·68	0·51	0·52
Et	1·22	0·97	—
Pr^i	—	1·30	1·40
Bu^t	—	1·68	—
COMe	0·01	−0·28	−0·37
OH	—	−0·49	−0·61
OMe	−0·33	0·12	−0·04
NO_2	−0·23	0·11	0·24

† Based on substituted phenoxyacetic acids. From Fujita, Iwasa, and Hansch (1964).

and for any extended range of compounds or substituents there are no clear relationships to ordinary polar or steric parameters.

For many biological activities, hydrophobic–lipophilic character by itself accounts for most of the observed structure–activity dependence. In such cases either the transport of the drug through membranes and biological fluids or the binding of the drug to the biological site is dominant. The binding of small molecules to biomacromolecules can sometimes be studied in isolation and is then found to depend on hydrophobic–lipophilic character. For instance, the binding of nineteen substituted phenols to bovine serum albumin is expressed by equation (136), with $r = 0{\cdot}96$, where C is the concentration of

$$\log(1/C) = 0{\cdot}68\pi + 3{\cdot}48 \tag{136}$$

phenol ($\text{mol}\,\text{l}^{-1}$) necessary to form a 1:1 complex with the protein. (For other examples see Cammarata and Rogers, 1972.)

In Table 18 are summarized correlations with P or π for various biological

TABLE 18

Correlation of biological activities with P or π†

Biological activity	*Equation*	*n*‡	*r*§
Inhibition of α-chymotrypsin by substituted phenols	$\log(1/K_i) = 0{\cdot}95 \log P - 1{\cdot}88$	10	0·99
Inhibition of *Avena* cell elongation by substituted phenylacetic acids	$\log(1/C) = 0{\cdot}73\pi + 3{\cdot}01$	18	0·97
Inhibition of guinea pig ileum contractility by *n*-alkanols	$\log(1/C) = 1{\cdot}06 \log P + 0{\cdot}62$	8	0·99
Narcotic action of miscellaneous aliphatic compounds on tadpoles	$\log(1/C) = 1{\cdot}17 \log P + 0{\cdot}68$	17	0·98

† From Cammarata and Rogers (1972).
‡ Number of compounds studied.
§ Correlation coefficient.

activities, with examples from enzymes, cells, tissues, and simple intact organisms. For a compilation of other examples see Cammarata and Rogers (1972); also the papers of Corwin Hansch, e.g. Hansch (1969).

In certain other systems the plot of $\log(1/C)$ against $\log P$ or π is parabolic, i.e. there is an optimum hydrophobic–lipophilic character. If the lipophilic character is too marked the molecules tend to be taken up by the first biomacromolecule they encounter and get no further. Such systems can often be correlated if terms in $(\log P)^2$ or π^2 are included. (Examples are in the above sources.) The procedure was originally highly empirical but Hansch (1969) has justified it in terms of a kinetic model.

There are, however, many systems for which consideration of electronic and/or steric effects seems essential.

The use of polar and steric parameters

We have already seen that the application of the Hammett equation to $v_{\max}$ and K_{m} in enzymology often gives rather poor correlations. There are few straightforward applications in pharmacology, mainly because of the dominance of hydrophobic–lipophilic character.

The LD_{50} values of fourteen *meta*- or *para*-substituted diethyl phenyl phosphates towards houseflies are given by (137), with $r = 0{\cdot}97$. A parallel study

$$\log(1/C) = 2{\cdot}28\sigma^{-} - 0{\cdot}35 \tag{137}$$

of fly-head cholinesterase inhibition by six of the *para*-substituted compounds yielded I_{50} values which were correlated by (138), with $r = 0{\cdot}98$. The similar

$$\log(1/C) = 2{\cdot}37\sigma^{-} + 4{\cdot}38 \tag{138}$$

ρ-values suggest that the toxicity of the compounds is a reflection of cholinesterase inhibition.

A related type of analysis uses pK_a values of drugs as polar parameters. Successful use of pK_a values can arise because at the pH value prevailing in the biological study, there are appreciable amounts of both ionized and unionized forms of drug present, but only the unionized form can enter the appropriate part of the biophase. There can then be a multiple correlation of biological activity with pK_a and the partition coefficient of the unionized form. For instance, the antibacterial activities (MIC) of fifteen arylamines towards *Escherichia coli* are expressed by (139), with $R = 0{\cdot}962$.

$$\log(1/C) = -0{\cdot}158pK_a + 0{\cdot}694\log P + 4{\cdot}462 \tag{139}$$

The Taft σ^* and E_s values find some application. The utility of E_s is remarkable in view of the systems being so very different from those used to define this parameter. A good example is provided by the use of σ^*, E_s^c (Hancock's modification, p. 46), and n_H (number of α-hydrogen atoms in an alkyl group) to correlate the anti-adrenaline effect (ED_{50}) of ten *N*-mono- and *NN*-di-alkyl-2-bromo-2-phenylethylamines on cat blood pressure,

$$\log(1/C) = 3{\cdot}56\sigma^* + 1{\cdot}11E_s^c - 4{\cdot}43n_H + 11{\cdot}91 \tag{140}$$

equation (140), with $R = 0{\cdot}98$.

The possible relevance of all types of parameter [hydrophobic–lipophilic (linear and square terms), polar, and steric] should always be considered for any given biological activity. The series of compounds should have numerous members, chosen to secure as wide a variation in each parameter as possible and avoiding too much multicolinearity of pairs of parameters (see Appendix). Unfortunately much pharmacological work is undertaken without this end in view.

Finally, there is a type of correlation analysis which employs only pharmacological data. This uses the 'additive statistical model'. Group contributions to the activity of a particular type of drug are derived and can be used to predict the activities of new compounds in the series [see Cammarata and Rogers (1972)].

PROBLEM

7.1. The Table gives LD_{50} values in the form $\log(1/C)$ for the toxicity of substituted benzoic acids towards mosquito larvae. Construct a table showing $\log(1/C)$, σ (from Table 1), and π (from Table 17) for the various compounds, and the π(BA) values given below, which are for the partition of benzoic acids (cf. the usual phenoxyacetic acid scale).

Substituent	$\log(1/C)$	π(BA)
4-I	2·31	1·14
4-Cl	2·06	0·87
4-Br	2·03	0·98
3-Cl	2·00	0·83
4-F	1·85	0·19
4-Me	1·66	0·42
H	1·64	0
4-OMe	1·60	0·08
4-NO_2	1·52	0·02
4-OH	1·29	−0·30

Plot $\log(1/C)$ variously against σ, π, and π(BA) and comment on the graphs.

The $\log(1/C)$ values of the 3,4-Cl_2- and 3,5-I_2 compounds are 2·28 and 2·85 respectively. Assuming additivity for the substituent parameters, insert points for these compounds on your graphs and comment. [π(BA) for 3-I is 1·28.]

Appendix: Introduction to statistical methods necessary for correlation analysis

THESE few pages are intended mainly to provide the necessary background in statistics for a proper appreciation of the subject matter of this book. They should also enable the reader to make a start on carrying out simple correlation calculations for himself. However, the account is no real substitute for a formal treatment of statistics, or at least sound guidance from a statistician. The books by Moroney (1956) and Snedecor (1946) will be found useful.

Simple linear correlation

In elementary physical chemistry if we have a series of values of a *dependent* variable y which we suspect may be related linearly to the *independent* variable x, we plot y against x. If the overall pattern of points seems to agree with the linear relationship, we endeavour to draw the 'best straight line' through the points. The equation of the straight line may be written as

$$y = ax + b \tag{141}$$

where a is the slope of the line and b is the intercept on the y axis. As long as the scatter of points about the line is commensurate with the estimated experimental error in y, we regard the linear relationship as well established. On the other hand, if the scatter seems out of all proportion to the likely experimental error, then we suppose that there is some complicating factor, possibly another independent variable which is influencing y.

It is not difficult to find research papers involving correlation analysis in organic chemistry in which the approach is essentially as just described. This is regrettable since the drawing of a best straight line by eye can be highly subjective, and the work usually lacks a proper assessment of the success of the correlation. The slope and intercept should always be calculated by the 'method of least squares', which readily leads on to proper estimates of success.

Before we deal with the method of least squares, two points of terminology must be mentioned. First, statisticians nowadays tend to use the term *explanatory variable* rather than independent variable. The term seems to have much to commend it and we shall make some use of it. A second point concerns our use of the term 'correlation' to describe the connection between the dependent variable and the explanatory variable. The statistician uses the term correlation specifically to refer to the strength of the association between two random variables in the population of the two variables. Where values of an explanatory variable are fixed in controlled experiments, the term *regression* is used. The latter situation seems to be the one with which we deal in correlation analysis in organic chemistry. We therefore make some use of the term regression in this book, while using the word correlation in its everyday meaning as a general description of what we are doing.

The method of least squares is based on the following convention: the best straight line is the line which estimates the values of the dependent variable y in such a way that the sum of the squares of the deviations between observed and estimated values is a minimum. In the development of the method it is assumed that the values of the variable, x, are precisely known and that the error is all in y. The necessity for this assumption may pose problems for us. In LFER there is often no reason to suppose that the $\log k$ (or K) values for one reaction series are much more precisely known than the values for the other. Thus the identification of one of our variables with y and the other with x may be difficult and arbitrary. Provided that the linear relationship is fairly well obeyed, it may not matter much what we choose to do. In any case, we are often concerned with a set of related correlations, and the main point is that we should act consistently. Thus in Hammett-type correlations we take the σ-constant as the variable x and $\log k$ (or K) as the variable y. (Inasmuch as the ionization constants of many benzoic acids are well established, this may well be justified for ordinary σ-values, but this would not necessarily be true for say, σ^+-values.)

If we have n pairs of values of y and x (data sets), then a and b are given by equation (142) and (143), which involve summations of the values of the variables, and of their squares and product. The simplest measures of the

$$a = (\Sigma xy - \Sigma x \Sigma y/n)[\Sigma x^2 - (\Sigma x)^2/n] \tag{142}$$

$$b = (\Sigma y - a\Sigma x)/n \tag{143}$$

success of the correlation are the *standard deviation of the estimate*, s, and the *correlation coefficient*, r. The significance of these is discussed below. They are given by the following expressions:

$$s = [\{\Sigma y^2 - (\Sigma y)^2/n - (\Sigma xy - \Sigma x \Sigma y/n)^2/(\Sigma x^2 - (\Sigma x)^2/n)\}/(n-2)]^{\frac{1}{2}} \tag{144}$$

$$r = (\Sigma xy - (\Sigma x \Sigma y)/n)/\{(\Sigma x^2 - (\Sigma x)^2/n)(\Sigma y^2 - (\Sigma y)^2/n)\}^{\frac{1}{2}} \tag{145}$$

Very few people any longer work out a, b, s, and r by using logarithms or a calculating machine. Numerous computer programs are available for this task and for the more difficult task of multiple regression (see p. 107). However, if you attempt an old-fashioned calculation, make sure you retain enough significant figures right through to the end. Premature rounding-off may lead to serious errors when you are dealing with small differences between large numbers. At the worst you may arrive at a straight line which does not pass between any of the experimental points, or a value of $r > 1$, which is absurd!

What is the significance of s and of r? The significance of s is that about two-thirds of the estimated values of y lie within s units of the corresponding observed value, while about 95 per cent of the estimated values lie within two standard deviations thereof. Almost never will an estimate have an error exceeding three standard deviations. If we are interested in the equation as a summary of the data, then the value of s is a means of assessing its reliability, and is also of use in considering the significance of deviant points. However, if we are interested in the regression as providing an 'explanation' of the results, one may reasonably argue that the scatter of the points, which is measured by s, should be looked at in relation to the numerical range of y covered by the data sets. A correlation for which $s = 0{\cdot}1$ units is highly meritorious if the values of y cover, say, 100 units, but is of little value if y itself covers only 0·2 units. These considerations are essentially embodied in the correlation coefficient, r, which is related to s by the equation:

$$r = [1-s^2(n-2)/\sigma_y^2 n]^{\frac{1}{2}} \tag{146}$$

where $\sigma_y^2 = \Sigma(y-\bar{y})^2/n$ and $\bar{y}$ is $\Sigma y/n$, i.e. the mean value of y. The quantity σ_y^2 is called the variance of the sample values of y. If $s = 0$, $r = 1{\cdot}000$ and this defines the perfect regression. As s increases, r decreases and becomes zero when $s^2(n-2) = \sigma_y^2 n$. This corresponds to there being no relationship between y and x, for there is then as much scatter from the line [measured by $s^2(n-2)$] as there is scatter from the mean value of y (measured by $\sigma_y^2 n$).

Since r is a square root function any value could be given a +ve or a −ve sign; r is conventionally taken as lying between 0 and +1 when a is positive, and 0 and −1 when a is negative. Computers commonly print out correlation coefficients according to this convention. However in correlation analysis in organic chemistry, the sign of r is generally omitted. Actually, the meaningful quantity is r^2, which gives the fraction of the variance of y which is 'explained' by the regression equation. Thus when $r = 0{\cdot}90$ the equation explains about 80 per cent of the variance.

Following Jaffé (1953) the calculation of correlation coefficients by physical organic chemists is a common practice. There is a widespread belief among chemists that values of correlation coefficients should be appraised as follows:

0·99 to 1·00	Excellent
0·95 to 0·99	Satisfactory
0·90 to 0·95	Fair
<0·90	Poor

although sometimes more exacting standards are applied. However, any such ideas are fallacious because correlation coefficients must always be considered in relation to the *number of data sets correlated*. The successive elimination of data sets will tend always to increase the value of r (unless sets showing substantial deviations are left in) and in the limit an apparently perfect correlation will be reached when $n = 2$, i.e. the minimum number of points necessary to establish the straight line! In fact a correlation of ten data sets giving $r = 0{\cdot}87$ (75 per cent of variance explained) is arguably a more successful correlation than one involving only four sets and giving $r = 0{\cdot}99$ (98 per cent of variance explained). The basis for saying this is that the *confidence level* or *significance level* for the former is 0·1 per cent but for the latter is 1 per cent. We must now see what this means.

These significance levels give the probability that the correlation could arise by chance from completely unrelated data. Thus 0·1 per cent corresponds to a probability of one in a thousand, while 1 per cent corresponds only to one in a hundred. Both these correlations are good, but the former is much better than the latter. Significance at the 25 per cent level means a probability of one in four that the correlation has arisen by chance, and this corresponds to a very poor correlation indeed. In correlation analysis in organic chemistry, any apparent correlation at a significance level worse than 5 per cent should be regarded with suspicion. We should usually expect better than 1 per cent. It should be noted that significance levels are sometimes expressed as the difference between 100 and the above percentages, e.g. the 0·1 per cent level is then called the 99·9 per cent level. (The convention adopted by any given author is usually obvious.)

Significance levels may easily be calculated from r and n by using 'Student's t function'. This is

$$t = r[(n-2)/(1-r^2)]^{\frac{1}{2}} \tag{147}$$

[Note that for a given r value, t increases with $(n-2)^{\frac{1}{2}}$.]

From the values of t and $n-2$ the corresponding significance level can be found from statistical tables (e.g. the Cambridge Elementary Statistical Tables) or from plots of t as a function of $n-2$ at various significance levels which are given in certain books [e.g. Moroney (1956)]. The quantity $n-2$ is sometimes designated as v, and is termed the *number of degrees of freedom*.

Thus for $n = 10$ and $r = 0{\cdot}87$, $t = 0{\cdot}87\sqrt{8}/\sqrt{(1-0{\cdot}87^2)} = 4{\cdot}99$. At $v = 8$, $t = 4{\cdot}50$ for a significance level of 0·2 per cent and $t = 5{\cdot}04$ for 0·1 per cent. The actual significance level is therefore very close to 0·1 per cent. Such a correlation is very good as an explanation of the data, but its performance as a summary, as measured by s, may well be poor. Thus one is often not justified in saying in absolute terms that a correlation is 'very good' or 'rather poor'. It all depends on whether one expects the correlation to provide an explanation or a summary.

One or two other matters regarding simple linear regression remain to be covered. In addition to providing values of a, b, s, r, and t, computerised least squares calculations often provide estimates of the standard errors of a and b. (It is also of course possible to calculate these by old-fashioned methods.) These are standard deviations and it is important to consider them when comparing a and b values for two regressions to see whether there is any significant difference between the respective a and b values.

Physical organic chemists have suggested various other quantities for expressing the precision of correlations. The main purpose in developing and using these is to get away from the misleadingly small numerical scale within which most values of r lie in physical organic chemistry. The simplest of these is essentially s/σ_y, which is written as SD/rms (by Taft) or ψ (by Exner). In a sense this is only a disguised correlation coefficient for it is equal to $[n(1-r^2)/(n-2)]^{\frac{1}{2}}$, but the numerical scale is expanded, and the factor of $n/(n-2)$ gives some consideration to the effect of degrees of freedom. Thus for $n = 10$, $r = 1{\cdot}00$ to $0{\cdot}90$ becomes $\psi = 0{\cdot}00$ to ca. $0{\cdot}49$, and correspondingly for $n = 4$, $\psi = 0{\cdot}00$ to ca. $0{\cdot}62$. It has been suggested that for many purposes values of ψ which are less than 0·10 may be considered to indicate a good correlation, while values between 0·1 and 0·2 indicate some relationship. It is claimed that ψ gives a clearer picture than r as to what a correlation really accomplishes. For instance, Exner (1972) gives a revealing example based on a Hammett correlation for which $r = 0{\cdot}956$ and $n = 12$. By most standards this would be deemed a satisfactory or even good correlation. Nevertheless $\psi = 0{\cdot}32$, which means that the measured values of $\log k$ are represented by the Hammett equation with a standard deviation of 32 per cent of that achieved by the simple assumption that substituents have no effect on reactivity!

Another possible measure of precision is to express the standard deviation of the estimate as a percentage of the range covered by the dependent variable y. Koppel and Palm (1972) do this for solvent effect correlations and use the symbol $s\%_0$.

It must be said, however, that neither of these measures of precision does anything which cannot be done by means of r, when properly interpreted.

Multiple linear correlation

The multiple linear regression of a dependent variable y with a series of explanatory variables x_1, x_2, x_3, etc. in equation (148) raises the question of

$$y = a_1x_1 + a_2x_2 + a_3x_3 \ldots + b \tag{148}$$

significance in an acute form. Most of our discussion will be about this.

There are rather complicated equations for applying the method of least squares to the regression of y on x_1 and x_2 and even more complicated equations for three explanatory variables. In correlation analysis in organic chemistry it is rare to undertake regression with more than three explanatory variables. Nowadays multiple regressions are almost invariably performed on a computer, so we will not waste space in stating the necessary equations. In addition to straightforward programmes for multiple regression, there are statistical analysis packages which permit the relevance of a large number of possible explanatory variables to be examined ('step-wise regression'). The computer will effect the regressions, starting with a linear regression on the most relevant explanatory variable, and then bringing in the other variables in the order of decreasing relevance, to some pre-determined level of significance. At each stage values of slope and intercept coefficients, and various statistical quantities, such as standard deviations and correlation coefficients, are calculated, so that the success of the correlation can be assessed. The question is 'What are the criteria of success?'

The basic statistical quantities are analogous to those of simple linear regression: they are the standard deviation of the estimate, s, and the multiple correlation coefficient, R. These are related by equation (149),

$$R = \left[1 - \frac{s^2(n-m)}{\sigma_y^2 n}\right]^{\frac{1}{2}} \tag{149}$$

where $(m-1)$ is the number of explanatory variables, and $n-m$ is thus the number of degrees of freedom, cf. the $n-2$ term in simple linear regression. The significance of R must be assessed in relation to n and m. Increasing the number of explanatory variables will increase R, but the apparent 'improvement' in any given case may or may not be significant.

Significance levels may be calculated from R, n, and m, by using the so-called F distribution. This is equation (150).

$$F = \frac{R^2(n-m)}{(1-R^2)(m-1)} \tag{150}$$

From the values of F, $m-1(=v_1)$, and $n-m(=v_2)$, the corresponding significance level can be found from statistical tables.

Thus for $R = 0{\cdot}93$, $n = 10$ and $m = 3$ (i.e. $v_1 = 2$ and $v_2 = 7$), $F = 22{\cdot}4$. At these values of v_1 and v_2, $F = 9{\cdot}55$ at the 1 per cent significance level and $F = 21{\cdot}69$ at the 0·1 per cent level. The actual significance level in this case is therefore very close to 0·1 per cent. As we said in connection with a similar situation in simple linear regression, the relationship in question may be regarded as providing a good explanation of the data (about 87 per cent of variance explained), but its performance as a summary, as measured by s, may be poor.

Possible improvement in a correlation by introducing a new explanatory variable must be properly assessed. This may be done in terms of a *partial correlation coefficient*. Let us take the use of a third explanatory variable as an example. For this purpose the dependent variable is usually designated 1, and the explanatory variables as 2, 3, and 4. The partial correlation coefficient, $r_{14\cdot23}$, for the introduction of variable 4 after 2 and 3 have been used, is given by equation (151), where $R_{1\cdot23}$ is the multiple correlation coefficient when 2

$$r_{14\cdot23}^2 = (R_{1\cdot234}^2 - R_{1\cdot23}^2)/(1 - R_{1\cdot23}^2) \tag{151}$$

and 3 are used, and $R_{1\cdot234}$ is the value when all three explanatory variables are used. The denominator gives the fraction of the sample variance of variable 1 which remains unexplained when variables 2 and 3 are used, and the numerator gives the additional fraction of the variance which is explained by introducing 4.

The significance level for the introduction of the additional variable may be evaluated from the t function, which in this case takes the form of equation (152),

$$t = r_{14\cdot 23}[(n-m)/(1-r_{14\cdot 23}^2)]^{\frac{1}{2}} \tag{152}$$

where $(m-1)$ is the number of explanatory variables, including the variable whose significance is being examined. Provided that the significance level is acceptable, say better than 5 per cent, partial correlation coefficients of as low as 0·7 may be of interest. If $R_{1\cdot 23}$ is 0·90, so that about 80 per cent of the variance is already explained, the explanation of 50 per cent *of what remains*, making 90 per cent in all, may certainly be of interest.

The quantities SD/rms $= \psi$ and $s\%$ (see above) have also been used to estimate the success of multiple correlations. [The relationship between ψ and R involves $(n-m)$.]

When several explanatory variables are used it is almost inevitable that there will be some degree of correlation between one 'independent' variable and another; this is called *multicolinearity*. Its effect is to increase the standard errors of the regression coefficients a_1, a_2, a_3, etc. [equation (148)]. Statistical analysis programmes commonly provide information about all this.

Conclusion

From what has been said it should be clear that valid correlation analysis depends on having an adequate number of data sets. Such sets must be well distributed over as wide a range of the variables as possible. When multiple regression is involved, multicolinearity must be minimized. These are counsels of perfection and cannot always be adhered to strictly but the limitations of supposed correlations should always be clearly stated. It is not difficult to find examples in the literature which are very deficient in this respect. In the space available in the present book it is unfortunately not possible always to deal adequately with statistical considerations.

Answers to problems

2.1. $\rho = 0{\cdot}924$, $\log k^0 = 0{\cdot}0368$, $r = 0{\cdot}994$, $s = 0{\cdot}0331$.

2.2 (*a*) The σ_I-type values follow the electronegativity order F > Cl > Br > 1. The σ_R-type values all show the numerical pattern F ≫ Cl > Br > I, corresponding to double bonding ability for the $-R$ effect. Cl and Br usually behave very similarly. The σ, σ^0, and σ_p^+ values show patterns arising from the appropriate combinations of σ_I-type (positive) and σ_R-type (negative) values, and sometimes depend on a quite delicate balance, e.g. the σ_m order F < Cl < Br > I depends on the competition of the electronegativity effect and the relayed resonance effect.
(*b*) *I* effect should follow the electronegativity order O > S; $-R$ effect should be O ≫ S [cf. (*a*)]. Relevant data for SMe from outside this book [see Exner (1972)] are: σ_m^0, 0·13; σ_p^0, 0·08; σ_p^+, $-0{\cdot}60$; σ_I, 0·19; σ_I(BA), 0·22; σ_R^0, $-0{\cdot}24$; σ_R^+, $-0{\cdot}85$; σ_R(BA), $-0{\cdot}24$.
(*c*) Such a group must be able both to release and to accept electrons by the *R* effect. The most obvious example of a $\pm R$ group is Ph. σ-values for this group are in Table 1; other LFER data for it include σ_p^+, $-0{\cdot}18$; σ_p^-, 0·11; σ_p^0, 0·04; σ_I, 0·10; σ_R^0, $-0{\cdot}10$; σ_R^+, $-0{\cdot}29$.

2.3. Most of the fifteen points clearly give a reasonable linear relationship. If little weight is attached to certain obviously deviant points such as COMe and CF_3, a line of slope ca. 1·13 may be drawn.

The fifteen substituents do not possess lone-pair electrons on the atom attached to the benzene ring. They are therefore incapable of a $-R$ effect. Some are clearly capable of a $+R$ effect, e.g. CN, NO_2, COMe, while others are clearly incapable, e.g. all CH_2X, SF_5. The situation regarding a possible $+R$ effect of certain substituents depends on the role attributed to d-orbitals of S, e.g. all SO_2X. The $-R$ substituents and the unipolar substituents in Table 1 deviate markedly from the line.

The main inference is that for the fifteen substituents the inductive effect operates more strongly from the *para*-position than from the *meta*-position by a factor of ca. 1·13. The $+R$ effect, where conceivable, apparently makes little contribution to the σ_p values. A plot of this type was used by Exner (1966) as his basis for separating inductive and resonance effects. The contribution of the $+R$ effect is greatly enhanced for σ_p^- values. See Table 3 and the following: CHO, 1·04; SO_2F, 1·32; SF_5, 0·70. For SO_2X groups a $+R$ contribution to σ_p^- is indicated, presumably involving participation of d orbitals, but not when the S is saturated, as in SF_5. As mentioned on p. 28, CF_3 appears capable of hyperconjugation in the sense of a $+R$ effect.

2.4. With σ, the plot is decidedly curved but is largely straightened out when σ^+ is used. With the latter, $\rho \sim 2{\cdot}0$. The $-R$ effect of substituents such as OH, OMe, and Hal clearly serves to stabilize the protonated form by cross-conjugation. (Write the canonical structures.) See Stewart, R. and Yates, K. (1958). *J. Am. chem. Soc.* **80**, 6355.

2.6. The ρ-values cannot be validly compared because one reaction is in water as solvent and the other is in benzene. A ρ-value for the ionization of substituted anilines in benzene is strictly required. In the absence of this, the best one can do is to look for ρ-values in solvents much less polar than water. van Bekkum, Verkade, and Wepster (1959) give $\rho = 4{\cdot}03$ in methanol, and 4·50 in ethanol. Clearly the ρ-value for ionization in any solvent as nonpolar as benzene would be very high. Thus the extent of C—N bond formation and the separation of the proton in the activated complex for the benzoylation reaction are not nearly so complete as the invalid comparison suggests.

2.7. The point for p-NH_2 is badly off the σ-plot for the other points, presumably because the two *ortho*-methyl groups twist the functional group out of the plane of the ring and thus inhibit conjugation involving ring and side-chain, cf. benzoic acid. σ^0 or σ^n is applicable, and σ^n seems to give the better plot with $\rho \sim 1{\cdot}7$, cf. 2·47 at 25°C in Table 6. However, we must correct the new ρ-value for the effect of temperature (p. 26). At 25°C, ρ would be about 2·3. The residual discrepancy, if significant, is probably due to solvent effects or to the influence of the two *ortho*-methyl groups on transmission of polar effects to the side-chain. The authors who obtained these rate constants misinterpreted the Hammett plot to indicate that the alkaline hydrolysis of methyl 4-X-2,6-dimethylbenzoates proceeded by an unusual mechanism. [See Goering, H. L., Rubin, T., and Newman, M. S. (1954). *J. Am. chem. Soc.* **76**, 787; cf. Bender, M. L. and Dewey, R. S. (1956). *J. Am. chem. Soc.* **78**, 317. To be fair, σ^0 and σ^n had not been invented in 1954. However, failure to allow for the effect of temperature was perhaps less excusable. For a general account of the mechanism of alkaline ester hydrolysis, see Ingold (1969).]

3.1. (*a*) The mean decremental factor per CH_2 group in the $CF_3[CH_2]_n$ groups $\sim 2{\cdot}8$, so σ^* for CF_3 should be about 2·6.
(*b*) The additivity value is $0{\cdot}215 + 0{\cdot}555 = 0{\cdot}77$.
(*c*) Me_3SiCH_2 is more electron-releasing than neopentyl. This fits in with electronegativities: C > Si.
9-Fluorenyl is significantly more electron-attracting than Ph_2CH. This is consistent with the ability of fluorene, as a cyclopentadiene derivative, readily to form a carbanion.

3.3 The E_s plot is easily the better, with $\delta = 1{\cdot}08$, cf. $\delta = 1{\cdot}00$ for acidic ester hydrolysis. Reactivity is governed almost entirely by steric effects; a small contribution from the polar effect is probably responsible for the deviations shown by $HalCH_2$. See Bolton, P. D. and Jackson, G. L. (1969). *Aust. J. Chem.* **22**, 527. Bolton has various other papers, mainly in *Aust. J. Chem.*, on acidic and also basic amide hydrolysis. When the number of α-hydrogen atoms is varied, Hancock's E_s^c and a hyperconjugative term are needed.

3.4. As long as R is of structural type XCH_2 and X is not very bulky, there is a good correlation with σ^*, and $\rho^* \sim -2{\cdot}6$, i.e. the rate is very sensitive to polar effects and is enhanced by groups which increase the electron density in the olefinic double bond. This is reasonable since molecular bromine is an electrophile.

When X is bulky or there is branching on the α-carbon atom, reactivity is reduced by a steric effect. The deviations are related to the E_s values in an approximately linear way. Now the role of steric effects is apparent, multiple regression on σ^* and E_s should be carried out for all sixteen data sets, but E_s values are not available for several of the substituents.

Ethylene and the polymethyl-substituted ethylenes can be considered in relation to the other compounds by attributing a change of σ^* of $+0{\cdot}49$ unit to the replacement of Me by H, and $-0{\cdot}49$ to the replacement of H by Me. The corresponding points deviate somewhat from the line specified above (ethylene below, the polymethyl compounds above) but quite a good new line for all the points could be drawn with a slope of ca. $-3{\cdot}0$. This is surprising, since replacing olefinic H atoms by methyl groups might have been expected to change steric influences seriously.

The rate data used in this problem are a selection from the extensive work of Professor Dubois and his colleagues; see, for example, Dubois, J.-E. and Mouvier, G. (1968). *Bull. Soc. chim. France*, 1426 et seq.

4.1. v against σ gives a fairly good straight line but there is considerable scatter. σ^+ and σ^- seem to possess no advantages for this system. (σ^+ puts the $-R$ *para*-substituents roughly on a straight line different from that defined by the *meta*-substituents.) The antisymmetrical stretching vibration of $—SO_2—$ is evidently insufficiently electron-demanding to enhance the $-R$ effects of *para*-substituents. Enhanced $+R$ effects would not really be expected, in view of the $+R$ nature of SO_2Cl. Taft's σ_o-values place the *ortho*-substituted compounds fairly well on the same plot as the *meta*- and *para*-derivatives, but *o*-Me deviates badly. No other set of σ_o values (Table 9) would do any better in this respect.

4.2. The ^1H shift shows an excellent linear relationship to the ^{13}C shift, with NH_2 deviating. In this particular system the factors governing ^1H and ^{13}C shifts seem closely related. The plot of the ^{13}C shift against σ_p shows much scatter, and the point for H is well off the best straight line. (σ_p^0 would not do any better.) The plot against σ_R^0 is much better. In this system the 4-^{13}C shift seems mainly governed by the resonance effect of the substituent, and this would apply also to the *p*-^1H shift, cf. the poor correlations with σ summarized in Table 11. The 4-^{13}C shifts show a reasonable linear relationship to the *p*-^{19}F shifts in substituted fluorobenzenes, so it is likely that multiple regression of the 4-^{13}C shift on σ_I and σ_R^0 would be effective (see p. 59).

4.3. The ionization is facilitated by electron accession to the bromine. There is a roughly linear relationship to σ^*, even including the point for HBr. The scatter of points is considerably larger than corresponds to the estimated error. The points for groups of the type R^1CH_2, i.e. with a constant number of α-hydrogen atoms, seem to show their own good linear relationship (although the range of I values is small). It is possible that a hyperconjugative effect of α-C—H bonds stabilizes the molecular ion, and that the stepwise removal of this effect in the series Me, Et, Pr^i, Bu^t opposes the influence of σ^*.

5.1. (*a*) $\varepsilon \sim 8$.
(*b*) and (*c*) are merely a matter of algebra. The equations are quite often encountered in these forms, although it is not always made clear that the equation in (*b*) involves an approximation, while that in (*c*) involves no approximation.

5.2. The points for ten solvents lie fairly well on a smooth curve (rather than a straight line), with diethyl ether, acetone, and methyl cyanide lying somewhat below the curve, and the alcohols showing gross deviations. Presumably the relationship between $\log k$ and E_T which is shown is due to the ion-pair-like nature of both the activated complex in the reaction and the initial state for the electronic transition (see p. 73). These are both stabilized by non-specific solvent–solute interactions and by electrophilic solvation. For the hydrogen-bonding solvents, however, the situation is apparently different. The alcohols stabilize the initial amine molecule through hydrogen bonding to the lone-pair electrons on the nitrogen atom. (See p. 74, and 77, and also Table 15 for detailed analysis of solvent effects on rate constants for this reaction and on E_T.)

6.1. The extent of charge delocalization in the anion of the acid catalyst is very important. Delocalization is much less in the anions from oximes and chloral hydrate than in the phenolate and carboxylate ions which define the line. On the other hand there is more extensive delocalization in the anions of aliphatic nitro-compounds and benzoylacetone. Charge delocalization influences K_a more than k_a, because in the transition state for the acid-catalysed reaction, the anion is not fully formed. Thus aliphatic nitro-compounds and benzoylacetone are worse catalysts and oximes and chloral hydrate are better catalysts than phenols of the same acid strength. [Draw the canonical structures of the various types of anion. For further details see Bell and Higginson (1949).]

6.2. There is clearly no relationship between $\log k$ and pK_a, but some relationship can be seen between $\log k$ and n or n_{Pt}. Thiourea deviates markedly from the n-plot but its high reactivity is well expressed by n_{Pt}. The SO sulphur atom is evidently a soft centre for attack by the nucleophiles. The data are from Kice, J. L. and Guaraldi, G. (1968). *J. Am. chem. Soc.* **90**, 4076. See also Kice, J. L., Kasperek, G. J., and Patterson, D. (1969). *J. Am. chem. Soc.* **91**, 5516 for data on the hydrolysis of an aryl α-disulphone, $ArSO_2{\cdot}SO_2Ar$. The SO_2 centre is much harder and the nucleophilic reactivity order is quite different, e.g. $F^- > Cl^- > Br^-$.

7.1. This problem well illustrates the difficulties of correlation analysis for biological effects. In all three plots a linear relationship can be discerned but there is considerable scatter of points and p-NO_2 deviates very badly from the Hammett plot. But for the p-NO_2 point, it would be difficult to say whether σ or a π-type parameter was the more successful. (This situation is due to a high degree of correlation between σ and the π-type parameters in the list when p-NO_2 is excluded.) The overall correlation is easily better with the π-type parameters, and a least squares procedure finds a higher correlation coefficient when π(BA) is used (0·946) than when π-values on the phenoxy-acetic acid scale are used (0·915). (Multiple regression on π-type parameters and σ values is no better.)

The points for the disubstituted benzoic acids lie rather below the line on the π-type parameter plots, indicating that the effective π values are less than those calculated on an additivity basis. In so far as π is characteristic of the whole molecule and not merely of the substituent, this finding is not surprising. On the Hammett plot, ignoring p-NO_2, the points for the disubstituted compounds lie fairly well on the line; σ-values often show good additivity.

[For data and other information see Hansch, C. and Fujita, T. (1964). *J. Amer. chem. Soc.* **86**, 1616; also Hansen, O. R. (1962). *Acta chem. Scand.* **16**, 1593.]

Bibliography

This Bibliography is highly selective. It aims to give information about books and review articles, either within the field of correlation analysis in organic chemistry or as background, together with references to key or 'classical' papers and to sources of data. It should provide the basis for further general reading in the field and for more specialized examination of the literature.

Reviews

The following books and articles are a selection of reviews in the field of correlation analysis in organic chemistry. Reference is made to some of them in several of the chapters in this book.

CHAPMAN, N. B. and SHORTER, J. (1972), Editors. *Advances in linear free energy relationships*. Plenum Press, London. References to individual chapters in this book are given below in the form: author's name (1972). *ALFER*. Chapter number.

CHARTON, M. (1971). The quantitative treatment of the ortho effect. *Progr. phys. org. Chem.* **8**, 235.

EHRENSON, S. (1964). Theoretical interpretation of the Hammett and derivative structure–energy relationships. *Progr. phys. org. Chem.* **2**, 195.

JAFFÉ, H. H. (1953). A re-examination of the Hammett equation. *Chem. Rev.* **53**, 191.

LEFFLER, J. E. and GRUNWALD, E. (1963). *Rates and equilibria of organic reactions*. Wiley, New York.

RITCHIE, C. D. and SAGER, W. F. (1964). An examination of structure–reactivity relationships. *Progr. phys. org. Chem.* **2**, 323.

SHORTER, J. (1969). Linear free energy relationships. *Chem. Brit.* **5**, 269.

——— (1970). The separation of polar, steric, and resonance effects in organic reactions by the use of linear free energy relationships. *Quart. Rev.* **24**, 433.

TAFT, R. W. (1956). Separation of polar, steric, and resonance effects in reactivity. Chapter 13 in *Steric effects in organic chemistry* (ed. M. S. Newman). Wiley, New York.

WELLS, P. R. (1963). Linear free energy relationships. *Chem. Rev.* **63**, 171.

——— (1968). *Linear free energy relationships*. Academic Press, London.

Chapter 1

The following books deal with the various aspects of reaction mechanisms, substituent effects, etc.

ALDER, R. W., BAKER, R., and BROWN, J. M. (1971). *Mechanism in organic chemistry*. Wiley, London.

HAMMETT, L. P. (1970). *Physical organic chemistry*, 2nd edn. McGraw-Hill, New York.

HINE, J. (1962). *Physical organic chemistry*. McGraw-Hill, New York.

INGOLD, C. K. (1969). *Structure and mechanism in organic chemistry*, 2nd edn. Bell, London.

JACKSON, R. A. (1972). *Mechanism: an introduction to the study of organic reactions*. Clarendon Press, Oxford.

KOSOWER, E. M. (1968). *An introduction to physical organic chemistry*. Wiley, New York.

SYKES, P. (1970). *A guidebook to mechanism in organic chemistry*, 3rd edn. Longman, London.

——— (1972). *The search for organic reaction pathways.* Longman, London.
WIBERG, K. B. (1963). *Physical organic chemistry.* Wiley, New York.

Papers referred to:

DAVIES, G. and EVANS, D. P. (1940). *J. chem. Soc.* 339.
DIPPY, J. F. J. (1939). *Chem. Rev.* **25**, 151.
FISCHER, A., MANN, B. R., and VAUGHAN, J. (1961). *J. chem. Soc.* 1093.
INGOLD, C. K., and NATHAN, W. S. (1936). *J. chem. Soc.* 222.

Chapter 2

BOLTON, P. D., FLEMING, K. A., and HALL, F. M. (1972). *J. Am. chem. Soc.* **94**, 1033.
BOWDEN, K., CHAPMAN, N. B., and SHORTER, J. (1964). *Can. J. Chem.* **42**, 1979.
BROWN, H. C. and OKAMOTO, Y. (1958). *J. Am. chem. Soc.* **80**, 4979.
BUCKLEY, A., CHAPMAN, N. B., DACK, M. R. J., SHORTER, J., and WALL, H. M. (1968). *J. chem. Soc.* (*B*) 631.
DEWAR, M. J. S. and GRISDALE, P. J. (1962). *J. Am. chem. Soc.* **84**, 3539, et seq.
EHRENSON, S., BROWNLEE, R. T. C., and TAFT, R. W. (1972). Paper in course of preparation.
EXNER, O. (1966). *Colln. Czech. chem. Commun.* **31**, 65.
——— (1972). *ALFER.* ch. 1 (a review of the Hammett equation).
HAMMETT, L. P. (1937). *J. Am. chem. Soc.* **59**, 96.
HEPLER, L. G. (1963). *J. Am. chem. Soc.* **85**, 3089.
KIRKWOOD, J. G. and WESTHEIMER, F. H. (1938). *J. chem. Phys.* **6**, 506, et seq.
MCDANIEL, D. H. and BROWN, H. C. (1958). *J. org. Chem.* **23**, 420.
SARMOUSAKIS, J. N. (1944). *J. chem. Phys.* **12**, 277.
TAFT, R. W. (1960). *J. phys. Chem.* **64**, 1805.
——— and LEWIS, I. C. (1958). *J. Am. chem. Soc.* **80**, 2436; (1959). *J. Am. chem. Soc.* **81**, 5343.
VAN BEKKUM, H., VERKADE, P. E., and WEPSTER, B. M. (1959). *Recl Trav. chim. Pays-Bas* **78**, 815.
YOSHIOKA, M., HAMAMOTO, K., and KUBOTA, T. (1962). *Bull. chem. Soc. Japan* **35**, 1723.
YUKAWA, Y. and TSUNO, Y. (1959). *Bull. chem. Soc. Japan* **32**, 971.
———, ——— and SAWADA, M. (1966). *Bull. chem. Soc. Japan* **39**, 2274.

Chapter 3

BOWDEN, K., CHAPMAN, N. B., and SHORTER, J. (1964). *J. chem. Soc.* 3370.
BRAUMAN, J. I. and BLAIR, L. K. (1969). *J. Am. chem. Soc.* **91**, 2126.
BROWN, T. L. (1959). *J. Am. chem. Soc.* **81**, 3229.
CHARTON, M. (1969). *J. Am. chem. Soc.* **91**, 6649.
HANCOCK, C. K., MEYERS, E. A., and YAGER, B. J. (1961). *J. Am. chem. Soc.* **83**, 4211.
MUNSON, M. S. B. (1965). *J. Am. chem. Soc.* **87**, 2332.
PAVELICH, W. A. and TAFT, R. W. (1957). *J. Am. chem. Soc.* **79**, 4935.
SHORTER, J. (1972). *ALFER.* ch. 2 (a review of the separation of polar, steric, and resonance effects).

Chapter 4

ADCOCK, W. and DEWAR, M. J. S. (1967). *J. Am. chem. Soc.* **89**, 379.
BROWNLEE, R. T. C., HUTCHINSON, R. E. J., KATRITZKY, A. R. TIDWELL, T. T., and TOPSOM, R. D. (1968). *J. Am. chem. Soc.* **90**, 1757.
BROWNLEE, R. T. C. and TAFT, R. W. (1970). *J. Am. chem. Soc.* **92**, 7007.
BURSEY, M. M. (1968). *Org. mass Spectrom.* **1**, 31.

——— (1972). *ALFER*. ch. 10 (a review of LFER and mass spectrometry).
FLEMING, I., and WILLIAMS, D. H. (1966). *Spectroscopic methods in organic chemistry*. McGraw-Hill, London.
KATRITZKY, A. R. and TOPSOM, R. D. (1972). *ALFER*. ch. 3 (a review of LFER and optical spectroscopy).
MCLAUCHLAN, K. A. (1972). *Magnetic resonance*. Clarendon Press, Oxford.
SCHMID, E. D. (1966). *Spectrochim. Acta* **22**, 1659.
TAFT, R. W., PRICE, E., FOX, I. R., LEWIS, I. C., ANDERSEN, K. K., and DAVIS, G. T. (1963). *J. Am. chem. Soc.* **85**, 709, 3146.
TRIBBLE, M. T. and TRAYNHAM, J. G. (1972). *ALFER*. ch. 4 (a review of linear correlations of substituent effects in n.m.r.).

Chapter 5

BROWNSTEIN, S. (1960). *Can. J. Chem.* **38**, 1590.
GRIMM, H. G., RUF, H. and WOLFF, H. (1931). *Z. physik. Chem.* **B13**, 301.
GRUNWALD, E. and WINSTEIN, S. (1948). *J. Am. chem. Soc.* **70**, 846.
HUGHES, E. D. and INGOLD, C. K. (1935). *J. chem. Soc.* 244.
KOPPEL, I. A. and PALM, V. A. (1972). *ALFER*. ch. 5 (a review of solvent effects on organic reactivity).
KOSOWER, E. M. (1958). *J. Am. chem. Soc.* **80**, 3253.
LAIDLER, K. J. (1963). *Reaction kinetics: Volume II—Reactions in solution*. Pergamon, Oxford.
REICHARDT, C. (1965). *Angew. Chemie* (*International Edn. in English*) **4**, 29.
SWAIN, C. G., MOSELY, R. B., and BOWN, D. E. (1955). *J. Am. chem. Soc.* **77**, 3731.
WINSTEIN, S., GRUNWALD, E., and JONES, H. W. (1951). *J. Am. chem. Soc.* **73**, 2700.

Chapter 6

BELL, R. P. (1941). *Acid-base catalysis*. Clarendon Press, Oxford.
——— (1959). *The proton in chemistry*. Methuen, London.
——— and HIGGINSON, W. C. E. (1949). *Proc. Roy. Soc.* **A197**, 141.
BRØNSTED, J. N. and PEDERSEN, K. J. (1924). *Z. physik. Chem.* **108**, 185.
EDWARDS, J. O. (1954). *J. Am. chem. Soc.* **76**, 1540.
EIGEN, M. (1964). *Angew. Chemie* (*International Edn. in English*) **3**, 1.
JENCKS, W. P. and CARRIULO, J. (1960). *J. Am. chem. Soc.* **82**, 1778.
PEARSON, R. G. (1968). *J. chem. Ed.* **45**, 581, 643.
PEARSON, R. G. (1972). *ALFER*. ch. 6 (a review of the role of the reagent).
——— SOBEL, H., and SONGSTAD, J. (1968). *J. Am. chem. Soc.* **90**, 319.
SWAIN, C. G. and SCOTT, C. B. (1953). *J. Am. chem. Soc.* **75**, 141.

Chapter 7

CAMMARATA, A. and ROGERS, K. S. (1972). *ALFER*. ch. 9 (a review of the interpretation of drug action through LFER).
DIXON, M. and WEBB, E. C. (1964). *The enzymes*, 2nd edn. Longmans, London.
FUJITA, T., IWASA, J., and HANSCH, C. (1964). *J. Am. chem. Soc.* **86**, 5175.
HANSCH, C. (1969). A quantitative approach to biochemical structure-activity relationships. *Acct. chem. Research* **2**, 232.
KIRSCH, J. (1972). *ALFER*. ch. 8 (a review of LFER in enzymology).
LAIDLER, K. J. (1958). *The chemical kinetics of enzyme action*. Clarendon Press, Oxford.
MICHAELIS, L. and MENTEN, M. L. (1913). *Biochem. Z.* **49**, 333.
NATH, R. L. and RYDON, H. N. (1954). *Biochem. J.* **57**, 1.

Appendix

MORONEY, M. J. (1956). *Facts from figures.* Penguin Books, Harmondsworth, England.

SNEDECOR, G. W. (1946). *Statistical methods*, 4th edn. Iowa State College Press, Ames.

Index

THIS index is a guide to the main entries for important topics. Individual compounds, reactions, etc. are not usually mentioned. T after the page number indicates a Table. Where a chapter is cited, page numbers within the chapter indicate particularly important matters.